I DARE YOU TO EAT IT

Liesa Card

I DARE YOU TO EAT IT

Designing Food Storage
You Would Actually Want to Eat

LIESA CARD

Clerestory 2008
www.idareyoutoeatit.com

ISBN 978-0-578-00187-6

For Samantha, Hailey, Sarah,
Rosemary, and Elizabeth.

With love, and in lieu of scrapbooks.

Acknowledgements

Probably the biggest mistake I made when I began this project was to think that I could just do it on my own. I had no idea it would be so hard. When it seemed impossible there was always a spark of hope and light that kept me moving forward.

In Luke, chapter 8 verses 2–3, there is mention of "certain women...which ministered unto him of their substance." I think I've known "certain women" (and men). Individuals who knew that following the Savior was the very best option and helped me do the same. My amazing mother, Marilyn Bennett Romney, three wonderful sisters—Linda Kirk, Carolyn Evans, and Virginia Hoffman—and the dearest of friends all helped me to see that women of substance choose to follow the Savior and dedicate both time and talents to His ministry.

The first person I told about my decision to write the book was Khaliel Kelly. Her ability to look food storage squarely in the face, seeing equally the humor and value, has encouraged me more than she'll ever know. The credit for the book title, website, and soon-to-be-released food storage doll belongs to Khaliel. Natalie Stillman-Webb taught me everything I still don't know about writing a book. A neighbor, YW buddy, visiting teaching sister, and friend, Natalie tirelessly worked to help me arrange the puzzle pieces of at least twenty-five years worth of story. Gently editing so close to my heart, Natalie did all the really hard work and never made me feel stupid. Stephanie Griffin lent a hand when I was at the end of my rope. A gifted writer herself, Stephanie gave her support as an early draft reader and talked me down from more than a few writing ledges. Beyond honest feedback,

Stephanie helped me draw out a bit more detail even when I felt completely empty. She helped me find the answers I needed.

When it comes to emotional support, my husband, Mike, is simply the best. At the beginning of the project he said, "Great." When I worked on it all through the night, he said, "Don't get sick." When I asked him to read early drafts he said, "This is going to be terrific." A man of few words, Mike has given me the love, encouragement, and space to just be who I am. When it came to the business of publishing the completed book, Mike stepped forward and led the way. He does all the driving on family vacations and I love him for it.

Our five daughters—Samantha, Hailey, Sarah, Rosemary, and Elizabeth—were raised in the teaching moment jungle. Always brave souls, they patiently endured and even appreciated the steps we took to follow the prophet. They are the real reason I worked so hard to understand, purchase, and practice using my food storage.

I DARE YOU TO EAT IT

Designing Food Storage
You Would Actually Want to Eat

Contents

Introduction
YOU CAN: Understanding the Potential of Provident Living   1

Chapter 1
I DID: Seeing Food Storage as Part of God's Plan   9

Chapter 2
HERE'S HOW: Planning to Eat and Share Your Food Storage   29

Chapter 3
START NOW: Implementing Your Food Storage Plan   55

Chapter 4
TRY THIS: Easy Food Storage Recipes   75

Chapter 5
GET GOING: Moving Beyond Food Storage Myths   129

Conclusion
TRUST: Preceding Testimony with Action   143

Resources
Cost and Weight Recommendations for Long-term
Storage Items   147

Food Storage Bulletin Boards for Meal Planning   149

Sample Three-Month Inventory and Shopping
List Spreadsheet   157

Introduction: Understanding the Potential of Provident Living

Investing in preparation

Think of one big accomplishment in your life: I'm sure it required some consistent effort to reach that goal, and no matter how difficult the challenge appeared, you found a way. It probably required some patience and a desire to learn new skills. Most of us are willing to work hard when we want to change and grow.

When I was thirty-five years old I applied to a design school in Manhattan. My husband, Mike, quit a good job to which he'd recently been promoted, and we moved across the country with our five daughters, living in a hotel for three months before we could find someone willing to rent to a large family. With twelve years of interior design projects behind me, my clients and even my friends thought I was nuts and argued that a formal education wasn't necessary or worth the sacrifice. But I believed that graduating from an excellent university would have value that reached beyond practical or obvious application. I believed in provident living.

For two years I felt crushed with the weight of endless assignments and design projects. In 1999 I graduated with honors and received one of the school's prestigious design awards. Beyond the academic adventure, New York was one example of the ways I've invested in the principle of provident living. It was a huge challenge for me but I overcame the obstacles. A college education furthered my personal preparation. I'll forever be grateful for both the opportunity and experience.

The Church of Jesus Christ of Latter-day Saints has taught me most of what I know about provident living. This concept has been defined as "careful in regard to your own interests; foreseeing wants and making provision to supply them; prudent in preparing for future exigencies" (http://dictionary.die.net/provident). The definition speaks of work and effort that improve future possibilities. I've come to understand provident living as living smart and armed for the future. We moved across the country because it was *smart* for both parents to have a college education.

And just as it is smart to have a diploma, it is smart to have food storage. We may not *have to* use either of them today, but preparation is always the more meaningful way of existence. Preparation enhances every opportunity.

Even so, along with college and other challenges in my life, food storage wasn't always something I was sure I could do. In trying to design what was best for my own family, the practical or "homework" part of the process bogged me down. Based on the recommendations from the Church, my family of seven needed to have about two thousand pounds of things like powdered milk, wheat, and dehydrated potatoes. That's a little scary. Like going back to school when I was too old, poorly dressed, with four young daughters and a nine-month-old baby, building a food storage appeared overwhelming.

Does it have to be weird?

I recognize now that I wasn't alone in my worries, that many of us have come to view food storage as so much scarier than it needs to be. It's a goal we'd like to accomplish, but the course appears too foggy and seems to fade into never-ending questions. How will we ever afford a year's supply of food? Where are we going to store it? How much are we supposed to buy? What do we buy? How do we cook all that stuff? What does it taste like? What if we really have to eat it? What do we think is going to happen?

Hold on. Seriously, what is the deal with food storage? Even animals know it's smart to store food, so why all the resistance? Why is it so easy to neglect this specific counsel from the prophet? It's like there's some force that really doesn't want us to live "careful in regard to [our] own interests; foreseeing wants and making provision to supply them." For some, food storage becomes a challenge that's simply overlooked or ignored.

But the concept of storing food for a rainy day has literally been around forever. And food storage is still being preached to us, over the pulpit, by a living prophet. In the November General Conference, 2001, President Gordon B. Hinckley stated, "As we have been continuously counseled for more than 60 years, let us have some food set aside that would sustain us for a time in case of need" ("The Times in Which We Live," *Ensign*, Nov. 2001, p. 73). It only makes sense that when the prophet counsels over thirteen million members worldwide to prepare food storage, there have to be ways for us to apply that counsel and benefit from it.

I found a way that works for my family

A main goal of this book is to help place the idea of foodstorage in context. First, we can begin viewing food storage as part of the gospel plan. So often, we tend to see it as a strange adjunct to the gospel, one that we can conveniently avoid; but instead, I'd like to discuss food storage as indeed a part of the gospel plan that involves faith, obedience, stewardship, and charity.

Second, we can see food storage as part of the context of our daily lives. It can be easy to view food storage as just one more thing we are supposed to be doing, on top of what may seem like already overwhelming commitments. Yet we may miss the potential that food storage has to streamline and actually simplify our lives. We already eat everyday and already spend a portion of our time planning and preparing meals: the system discussed in this book places food storage within the context of the individual family's routines, tastes, and budget. Instead of buying, storing, and then tossing it, we can create a food storage that is completely integrated into daily life.

The Prophet Spencer W. Kimball emphasized these practical concepts when he taught, "Preparedness, when properly pursued, is a way of life, not a sudden, spectacular program. We could refer to all of the components of personal and family preparedness, not in relation to holocaust or disaster, but in cultivating a life-style that is on a day-to-day basis its own reward" (*Teachings Of Presidents Of The Church: Spencer W. Kimball*, Salt Lake City, UT: The Church of Jesus Christ of Latter-day Saints, 2006, p. 121–22).

I finally found the course that works for my family: it involved 1) gradually building a year's supply of basic dry goods suggested by the Church, 2) designing thirty meals that used these dry goods, and 3) purchasing a three-month supply of the canned, dried, or bottled ingredients for these thirty meals, with fresh ingredients added as necessary and as available. Now we have food storage we're comfortable eating on a daily basis.

I'd like to share some basic strategies with you

The chapters that follow offer my personal experience—successes and failures—with food storage. I've shared the insights I gained during twenty-five years of struggling to be obedient, as well as the shift in understanding that ultimately enabled me to bridge the gap between building a food storage and actually eating it. There are specific techniques for setting yourself up for success, plus examples of easy food storage recipes, followed by one rather large pep talk.

This book is about offering guidance that will walk you through designing and putting food storage into place *your own way*. Investing in the design process for your own food storage will eliminate the fear of using it, since you can create a storage that makes sense and fits your life. I hope that as you read you'll consider what the prophet has suggested, what your family might want, and how you can help others.

I'm not going to try to cover everything there is to know about food storage, first of all because I don't know it all and second because I don't have time. What I do know is how to eat my food storage, and I don't need hundreds of pages to explain the method. This book is about a commitment to following the counsel from our prophets, developing a strategy to better provide for our families and friends, and then getting on with life as it is right now.

In his first published article as sixteenth Prophet of The Church of Jesus Christ of Latter-day Saints, President Thomas S. Monson counseled, "Learn from the past. Prepare for the future. Live in the present" ("Treasure of Eternal Value," *Ensign*, April 2008, p. 4). I believe that food storage can help each one of us accomplish those directives. Surely this isn't the most difficult thing we've been asked to do. In fact, within the process of designing and using our food storage we can eventually make life easier, every single day.

If I can figure it out, you can figure it out

I must admit up front that cooking has not been my career or hobby. In fact, I'm a perfect example that no one has to be a cooking expert in order to successfully build and use food storage. Even now I don't especially love to cook and it doesn't come naturally to me. I admit to using a recipe when I boil eggs. Laugh if you must. Only in the last couple of years have I dared to wing it. But provident living is the prophet's "come as you are" party and we've all been invited. Our individual challenge comes in discovering how to use the gift and make our way to the blessings of preparedness. I would like to help with that process.

(Figuring things out is my job)

Maybe the one edge I have, and it's a very small edge, goes back to my twenty plus years of working as an interior designer. In addition to finding good homes for expensive sofas, I ultimately get paid every day to view things from a different angle. Each job is a unique puzzle, and success in my business comes when I am able to design something that uniquely suits the people I am working with. Developing my food storage plan has been a design project in which I've worked to understand what suits my life and my family.

Once I figured out how I could design and use my food storage all the time there was an actual shift in my lifestyle. I was prepared with the things my family was used to eating, I knew exactly how to cook with the supplies I had stored, and I began reaping the numerous blessings of that preparation. No earthquake required.

And I have a testimony of food storage

Using a business approach and lots of faith, I eventually was blessed to see food storage "differently" and then designed an expanded program founded on the inspired recommendations from The Church of Jesus Christ of Latter-day Saints. Designing thirty dinners that use staples of wheat, rice, beans, pasta, and potatoes in combination with everyday grocery store ingredients has become my strategy for overcoming the challenges of food storage. I've experienced for myself how the principles of provident living have streamlined the meals I serve, and even though cooking isn't my deal, I've been able to get organized and prepared. This approach to food storage fits my lifestyle as a working mother, suits my family's preferences, and holds the potential for serving others. It slices! It dices! No, not really. We're still just talking about mostly canned goods. By far, the best thing about my food storage is that after years of trying to obey the prophet's counsel on temporal preparedness, I was led to some very specific pieces of inspiration. This book is really about that story.

1

I DID

Seeing Food Storage as Part of God's Plan

My food storage journey began in my parents' home

Stuff happens

When I was a teenager, my father's business failed and then about a year later he suffered a heart attack. Thankfully he was blessed to survive both catastrophes. My parents did everything in their power to continue to shield and provide for their family as best they could, and having a little food stored in the basement was a blessing. My mother didn't have an organized system, fancy shelving, or even the room for food storage in a nine-hundred square-foot house, but somehow she was able to squirrel away cans of wheat, rice, beans, flour, sugar, and honey. I still remember how wonderful it was to be able to open her jars of pears, peaches, and those amazing cherries that she bottled every summer.

So at fourteen I learned that reversals could happen in regular, hard-working, faithful families, and that having even a little extra food set aside would prove to be helpful. On paper, my parents couldn't afford food storage. Six children and owning their own business meant they were always trying to stretch the budget in order to "make ends meet." When I complained—and at that age I always complained—at having to help prepare for the annual bottling of what seemed to be endless boxes of fruit, my sweet mother calmly insisted that we do this simply because the prophets had asked us to. She never tried to justify or even convince. She just lived a humble life that exemplified obedience. She had resolved that regardless of income she would do all in her power to obey the prophet's counsel. I love my mom for doing that.

Having parents who live the gospel helps

My good parents took us to church, taught us about the scriptures in our home, and then worked hard to live the principles they were preaching. More than their words, I remember what I saw them do. Whether food storage was mentioned in General Conference or our neighborhood ward, I witnessed my mother, with the slightly nervous support of my father, always re-examining her situation and searching out new ways to obey. She must have postponed certain purchases or found something else we could do without, because every so often a few more cans would appear in the basement. Food storage didn't easily "fit" into our budget but she jammed it in there anyway. For her, the spirit and letter of the law were the same thing, and she relied on the Spirit to help her accomplish the letter. My parents were never wealthy in terms of the world and yet their gift of faithful example was priceless.

They showed me what it means to be a courageous follower

Looking back, I can see that gift of obedience and faith was the dowry I took into my own marriage. From the very beginning, when Mike and I listened to our leaders and prophets speak about food storage we assumed they were talking to *us*. We never saw ourselves as exempt because we were young and going to school. We thought we were meant to try to follow their counsel. We believed we were accountable for the information—that it was truly given in love, and would absolutely lead to blessings.

I began to buy food storage about two years into our marriage, when we had our first child. Amid everything young parents do to prepare for a new baby, I started to consider my ability to always provide for the necessities of her life. I began thinking about preparing for my child in the ways my mother had prepared for me. I wanted to provide my child with an excellent education, comfortable home, nice clothes, dancing lessons, soccer tournaments, and piano recitals, but so much more important than all of that, I wanted to always be able to give her what she would need to stay alive.

It still wasn't easy and came down to
sorting through my family's wants and needs

With little more than ears that could hear and a fundamental desire to be prepared, I simply began buying two cans of soup instead of just one. It's where my obedience in food storage began. Within eighteen months of Samantha's birth we were blessed with twins, Hailey and Sarah. Mike was working forty hours a week at the front desk of the Union Plaza Hotel while taking sometimes as many as twenty-one credit hours per semester at the University of Nevada, Las Vegas. Those were some wacky years. I've suppressed a lot. With three children in eighteen months, a full-time job, full-time school, and not a single relative in sight, both of us were working really hard all the time.

We didn't have extra money. Some years we asked for food storage as our Christmas or birthday gifts from Mike's parents. Sometimes we even used small portions of our student loans to pay for it. Nobody panic. We didn't go crazy with debt but actively looked for things, like cable TV and expensive clothes, that we could go without in order to buy food storage. One time, we even piled bags of wheat in the kids' bathtub.

I recently found an old receipt from a food storage company dated 12-16-88, listing items such as powdered milk, white rice, brown rice, and beans, for a total of $317.80. We had been married for about five years, with a 3½ year-old daughter, 2-year-old twins, and a baby due in 6 months. Those were the kamikaze child bearing years so it's amazing to me that we were focusing on food storage then, but we overcame our own storage objections and found a way. The love we felt for our daughters, combined with faith in a living prophet, made us take action when he spoke about food. We made a decision to be compliant. That's the good news.

Buying food storage was difficult,
but cooking with it was an even bigger challenge
The bad news is that years down the road and despite my testimony of obedience, I still was never very good at *using* our food storage. I faithfully listened to Relief Society lessons, set goals, bravely experimented with new recipes, and then went back to my old ways of cooking for my family. No matter how hard I tried, I always seemed to fail when it came to using our food storage.

How could I incorporate food storage into my daily life?
We had tried to be obedient, we had spent a significant amount of cash, we had hauled all those cases of food across the country a couple of times, and then it sat piled in the basement. I was always too busy with a marriage, five daughters, extended family, a church calling, and an interior design business. Preparing meals from my food storage while I was doing my best to survive "normal" life seemed much too complicated.

And yet, I just knew that the prophet's counsel was intended to be more useful in my day-to-day life. It bothered me that the supply of food we had worked and sacrificed for was only taking up space and growing old. Nutty as it sounds, I wanted to buy more! I wanted to expand into the foods that we regularly ate but couldn't justify further expenditures until I was actually using what we already had. More prayer was needed.

And then an answer came
All that collective time and effort spent trying to get my brain, and budget, around something as large as a year's supply of food was eventually rewarded with an actual flash of enlightenment. Maybe the blessing came because I truly had gone as far as I could on my own. After years of doing my best to obey the inspired counsel to have a basic food storage, I was blessed with a very specific bit of personal inspiration that came in the form of a sketch.

Seeing the whole food storage picture
Sharing personal inspiration is extremely tricky. Besides the fact that the information is sacred, it's almost

impossible to accurately describe using mere words. It's a unique form of communication that comes into our hearts and then enlightens our minds. Inspiration usually comes to me in the early morning hours. It's the time when my world is the most still and I can review questions in my head, ponder ideas in my heart, and search for answers. One morning, following personal scripture study and prayer, a whole new perspective on food storage suddenly came to me.

I saw a circle cut into three equal, pie-shaped sections. The top section was labeled "Prophet," the left section was labeled "Us," and the right section was labeled "Them." The Holy Ghost truly sent me a drawing!

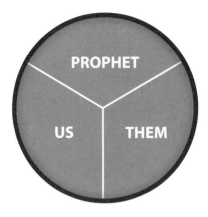

Really—I could see this in my head

The drawing was simple and precise, and it illustrated to me how food storage ties into Heavenly Father's larger plan for his children. All three sections—"Prophet," "Us," and "Them"—were exactly equal, depicting that the words of the prophet matter, one family's needs matter, and the needs of everyone else matter, too. Beyond some politically correct notion, I could see the diagram as tangible evidence that everyone really is equal in value and everyone can benefit from this inspired program.

As well as I can explain, the circle shape implies connection: each section of the drawing touches the other two sections, just as the prophet, us, and everyone else are all connected and linked. A circle also symbolizes movement and progression. We're in this together. And the circle illustrates that food storage is much bigger than preparing for the next earthquake or job layoff but is part of the whole gospel plan.

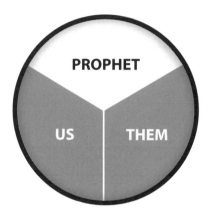

The Prophet

In the drawing, the prophet's position at the top center section signifies that everything stemming from the prophet matters. He is the source. His direction is the highest form of information and it comes first. The "Prophet" portion also stretches across the "Us" and "Them" sections like a giant umbrella. Prophetic leadership and counsel offer coverage that can protect everyone. When storms happen, real protection comes when we have chosen to stay connected to the prophet and to each other.

Also, the prophet is the prophet for the whole world. Not just for LDS people. The real revelation he gives at General Conference isn't just for "us." Samuel the Lamanite wasn't just talking to a select group when he gave his message that Christ was coming. Same thing with Isaiah or Paul. Their messages were for everyone, not just Jews or Christians or the faithful. Messages from our Heavenly Father come through the prophet to bless and direct all of God's children. So, even when the prophet speaks about food storage he's really talking about something that can bless everyone—"us" and "them."

It's smart to follow a prophet

In this flash of inspiration the Spirit was encouraging me to continue to trust in direction from a living prophet. "Remember that He speaks for Jesus Christ" kept filling my heart and mind. That was huge to me because there had been numerous moments of insecurity and doubt during all those years of buying mostly dry goods. Dry goods aren't especially appealing and sometimes I felt a little embarrassed about my purist stash. Clearly I wasn't successful at using it, and so I questioned whether my storage was too basic, too hard-core. Maybe I was buying the wrong stuff!

I was comforted with the realization that the prophet knows all about Costco and is still counseling us to buy the food "required to keep [us] alive if [we] did not have anything else to eat" (First Presidency Letter, *Ensign*, March 2006, p. 70). I felt grateful for the simple guidelines from the Church. I don't think the guidelines are weird. Are they basic? Thankfully, yes. Are they inspired? Again, yes—I'm not going to kid myself into thinking I'm smarter than the prophet but will choose to follow his direction.

He knows all about the cans
stashed under beds and piled in closets

Take a look at what President Gordon B. Hinckley stated in his Relief Society Conference address of October 2006:

> Today membership in the Relief Society is somewhere around five million[...]Relief Society stands for self-reliance. The best food storage is not in welfare grain

elevators but in sealed cans and bottles in the homes of our people. What a gratifying thing it is to see cans of wheat and rice and beans under the beds or in the pantries of women who have taken welfare responsibility into their own hands. Such food may not be tasty, but it will be nourishing if it has to be used. ("In the Arms of His Love," *Ensign,* Nov. 2006, p. 116)

When I heard the prophet's statement I wanted to stand on my chair and clap. I'm so thankful for his direct acknowledgement that while food storage may not be "tasty" we should go ahead and *buy it anyway.* His message today is less about fear and warning and more about forethought and determination.

Food storage is the prophet's insurance plan that offers coverage and protection to everyone. We are major players in the prophet's plan. Without the participation of individual families the plan won't work. The Church organization has many wonderful, large-scale programs in place to assist regions in natural catastrophes or individuals in economic straits; but all of that can't discern and address every need the way that neighbors in a community can. By prayerfully implementing food storage we become part of the solution in any crisis, whether it be a flood or a layoff. Following the prophet's counsel to have food storage empowers the individual to act, to immediately reach out, to lift and bless. That's what the Savior did during his earthly ministry.

And boring food storage is really valuable

I felt great comfort and confirmation that those early steps we had taken as a young family (in virtually blind obedience) to purchase wheat, rice, beans, and oats was a great place to begin. The "boring food storage" was relatively cheap and yet highly durable, with the greatest potential for lasting the longest and serving the most people. Our finances didn't allow us to move very fast in the beginning, and our supply wasn't being rotated as it should, but we had made a start and it represented our obedience.

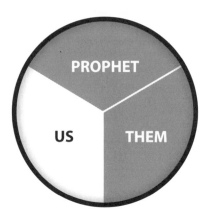

Us

At the same time that we have this broad, global, every-one-matters view, food storage is also meant to answer the specific needs of one family. Each family's needs matter. We strive to follow the counsel of the prophet and our Heavenly Father, but then we must own our food storage. We need to invest ourselves, make it our stewardship, and make a plan for our family.

My family's wants and needs matter

As I considered the section of the drawing labeled "Us," I realized that this is where my food storage preparations needed to grow. While it was important to have a deep supply of basic dry goods, I also needed to include items my family eats day to day. Simply having 500 pounds of wheat stored isn't enough: the essentials matter, but then we each have to figure out how to tailor our food storage to best fit our needs.

Another line from the First Presidency letter (*Ensign*, March 2006, p. 70) emphasizes that food storage is meant to be customized: "When members have stored enough of these essentials to meet the needs of their family, for one year, they may decide to add other items that they are accustomed to using day to day." Most recently the First Presidency has published a pamphlet titled "All Is Safely Gathered In: Family Home Storage" (Intellectual Reserve, Inc, 2007): the first thing it suggests is that each member store a three-month supply of the *foods we are accustomed to eating on a daily basis.*

I needed to design meals that didn't scare the children

I realized I needed to plan my family's meals as if I were running a restaurant. Using my long-term storage items as a base, I would choose recipes that combined the prophet's basic suggested products with the more familiar ingredients we normally use from the grocery store. More typical foods weren't going to *replace* what the prophet had suggested but would simply be an addition. The #10-sized cans in the basement would be opened and the dry goods would get used up because of recipes that incorporated them into our normal, daily diet.

For us, those typical foods include plenty of instant, or almost instant, options. If a few goofy cans of chili, for example, mean I'm a tiny bit more prepared for one more family dinner, I'm good with that. I need serious help getting meals on the table. But despite a too-hectic schedule I believe it's important to continue the practice of family dinners. In the event of an emergency I could serve the canned chili over a bed of rice, or wheat, or plain beans, and hope that others would be good with that.

Beyond marking each item with its purchase date and having great shelving systems that neatly roll cans forward like homemade vending machines, the real truth about rotating food storage is that you *eat it*. That's how we're supposed to use it "day to day." Moving it around on the shelves is not the whole story. The ultimate goal is to eat the food and then buy more food. It's that very process of eating and replacing which keeps food storage fresh and useable. And if my family enjoys the ways I prepare my food storage there isn't any resistance to eating it.

Them

Our neighbors matter, too, in this conception of food storage. They are next door and right beside us. Their part in this drawing and in our Heavenly Father's plan is vital and significant. And every single one of them can have full access to the prophet, encircled in the love of the Savior. Love isn't based on membership or finances and neither are the blessings of food storage. Mormons haven't been given counsel to prepare food storage because they're "chosen" or because the Lord wants to save only LDS families. Like the ancient Jews, we've been given guidance that is not just for "us." The guidance we are blessed to receive from our leaders has work and responsibility attached to it.

As Paul explained in Romans 3:1–2 (*Bible*, King James Version), "What advantage then hath the Jew [over the Gentile]?[...]Much every way: chiefly, because that unto them were committed the oracles of God." The "oracles" Paul referred to are revelations we know as scripture. What has come to be our Bible came through Judah and was for everyone. Members of the Church have a responsibility to take what comes through modern prophets and share it with others.

I need to remember my neighbors—
I need to remember "them"

The prophet's plan for food storage is today—as it has always been—a plan for everyone. As the First Presidency recently noted (again in the "All Is Safely Gathered In" pamphlet), "prepare every needful thing" (see D&C 109:8), so that, should adversity come, we may care for ourselves and our neighbors and support bishops as they care for others." If our bishops send out a call for food, who will be prepared to serve? Who will be ready to accept that calling?

Asking myself those questions, I knew I wanted to help others but was very aware of the limited amount I could store. Constantly pressed by a budget, I couldn't help questioning how there could ever be enough to serve those beyond my own family.

I was reminded that the Lord knows how to stretch food

Powerful scriptural examples of obedience and faith that prompted miracles and extended supplies of food in order to bless others came back to me. In the Old Testament, 1 Kings 17:9–24, the impoverished widow of Zarephath and her son answered a prophet's call for food. When Elijah asked, this woman facing starvation chose to let go of fear, trust in the Lord, and contribute all that she had. Her faith unlocked incredible miracles:

And she went and did according to the saying of Elijah: and she, and he, and her house [child] did eat many days.

And the barrel of meal wasted not, neither did the cruse of oil fail, according to the word of the Lord, which he spake by Elijah. (Kings 17:15–16)

A prophet's life was sustained, which enabled him to continue teaching the word of God. A destitute single parent had enough food to sustain herself and her son. And certainly the most significant miracle for the widow of Zarephath, Elijah later raised her son from the dead. Here's a woman who committed her last morsel of food to the kingdom and was ultimately rewarded with restoration of her child's life. Could I do all that she did? I have faith that, as I obey a prophet's call for food and trust inspired leadership, when it comes time to share it my children will be blessed in ways they need most and as only the Lord knows.

In the New Testament, Matthew 14:17–21, "five thousand men, beside women and children" were fed from only five loaves and two fishes. A day of endless compassion, after all the sick were healed, culminated in another opportunity where limited resources were miraculously stretched. The Savior could have sent the "great multitude" away to obtain their own food in the villages as his disciples thought necessary, but instead chose to *extend* the small amount available for sharing. Jesus is just as kind and generous today as he was back then. He still understands our needs. We can thrill to consider that His blessing may expand our humble offerings, that all may be "filled".

Sharing, not hoarding

The Lord works with the faith and obedience of His followers in unlocking potential miracles. Counsel from a living prophet is a blessing; choosing to listen to that counsel and storing extra food gives the follower an opportunity to share and bless others. We become partners in an eternal work. Surely the Savior is planning that food storage be shared and that we have a part in the miracle.

One of my favorite stories from the New Testament is found at the end of the four gospels, in the last chapter of John, chapter 21, when Jesus appears to the disciples at the sea of Tiberias and teaches them to "Feed my sheep." This section of the Bible is often quoted in connection to missionary work, but is it possible that the Savior was speaking of nourishment literally as well as metaphorically?

A group of disciples, including Peter and John, embarked on an all-night fishing trip and caught nothing.

> But when the morning was now come, Jesus stood on the shore: but the disciples knew not that it was Jesus.
>
> Then Jesus saith unto them, Children, have ye any meat? They answered him, No.
>
> And he said unto them, Cast the net on the right side of the ship, and ye shall find. They cast

therefore, and now they were not able to draw it for the multitude of fishes.

Therefore that disciple whom Jesus loved saith unto Peter, It is the Lord. Now when Simon Peter heard that it was the Lord, he girt his fisher's coat unto him, (for he was naked,) and did cast himself into the sea. (John 21:4–7)

John sees the nets almost breaking from the overwhelming catch, instantly recognizing the Savior's work, and Peter has jumped out of the boat and is in the water, swimming for all he's worth. His love and commitment to the Savior is almost reckless. I want to become more like Peter! I hope if I ever have the chance to see Jesus I will *run* to be near Him and then kneel at his feet.

The best part of the story comes a few verses later, after the resurrected Lord and his disciples have shared a meal from the miraculously abundant supply of fish:

So when they had dined, Jesus saith to Simon Peter, Simon, son of Jonas, lovest thou me more than these? He saith unto him, Yea, Lord; thou knowest that I love thee. He saith unto him, Feed my lambs.

He saith to him again the second time, Simon, son of Jonas, lovest thou me? Yea, Lord; thou knowest that I love thee. He saith unto him, Feed my sheep.

He saith unto him the third time, Simon, son of Jonas, lovest thou me? Peter was grieved because he said unto him the third time, Lovest thou me?

> And he said unto him Lord, thou knowest all
> things; thou knowest that I love thee. Jesus saith
> unto him, feed my sheep. (John 21:15–17)

Again, and again, and again, the Lord commands Peter
to feed his sheep because he *really wants him to understand
this message. Feeding and caring for people is exactly what
love for the Savior looks like.* If this feeding is limited to
a metaphorical interpretation then we have a wonderful
scripture about missionary work, which is certainly valid
and inspiring. But just as we interpret the healing of the
ten lepers as literal maybe we can also read this "feeding"
as literal. The Savior's healings were certainly literal,
so it makes sense that his miracles in conjunction with
food—where he turned scarcity to abundance—not be
solely limited to symbolism. I think the Savior seriously
cares about our temporal needs, and with food storage
comes the opportunity to help with that blessed work.

2

HERE'S HOW

Planning to Eat and Share Your Food Storage

Understanding the scope of a food storage program as part of God's plan, we still have to develop a method for reaching specific goals. This next chapter will help you move from lovely ideas to application and functionality.

Planning that follows divine guidance: look at what the prophet has suggested

Let's start at the top: food storage implementation means listening to the prophet and then doing all we can to obey. It means putting into practice what is believed to be divine guidance.

The latest direction from the First Presidency ("All Is Safely Gathered in: Family Home Storage," Intellectual Reserve, Inc., 2007) is to "build a small supply of food that is part of your normal, daily diet." It suggests we "store drinking water for circumstances in which the water supply may be polluted or disrupted" and "establish a financial reserve by saving a little money each week and gradually increasing it to a reasonable amount." There is also counsel to "gradually complete your one-year supply with food that will last a long time and that you can use to stay alive."

Luckily for me and everyone else, The Church of Jesus Christ of Latter-day Saints has created an online food storage resource: www.providentliving.org. Learning for the first time, or re-learning the principles that have been taught for years, we are offered an efficient way to access the suggestions from inspired leaders. I appreciate that the information found online is practical and concise. Talks from President Gordon B. Hinckley, President Thomas S. Monson, Elder L. Tom Perry, and Bishop Keith B. McMullin are quick reads and powerful reminders of counsel we've received in General Conferences.

Solid answers to food storage questions eliminate rumors that sometimes lurk around the topic of preparedness. Information includes basic recipes, a short list of suggested long-term storage items, a handy food storage calculator, suggestions about packaging, and locations for home storage centers in different parts of the country.

Be sure to read the Frequently Asked Questions on the Church's Provident Living web site

My favorite is the first question, because it gets to the heart of food storage issues and answers the biggest question of all:

What is the most important thing I can do regarding family home storage?

Get started! If you have already begun, faithfully continue your efforts. As President Hinckley taught: "We can begin ever so modestly. We can begin with one week's food supply and gradually build it to a month and then to three months. I am speaking now of food to cover basic needs. As all of you recognize, this counsel is not new. But I fear that so many feel that a long-term food supply is so far beyond their reach that they make no effort at all. Begin in a small way, my brethren, and gradually build toward a reasonable objective" (In Conference Report, Oct. 2002, p. 65; or *Ensign*, Nov. 2002, p. 58). (www.providentliving.org)

I love being reminded that our leaders' number one message on food storage is to *get started* on the basics! I love the suggestion that food storage doesn't have to be overwhelming and the encouragement to stop being the people who make "no effort at all."

Another practical resource is the Family Home Storage Center

In the Family Home Storage Center, the Church has provided us an excellent resource for beginning our efforts. Commonly known as dry-pack canneries, these facilities offer an inexpensive and reliable way to build a long-term storage of dry goods. Certainly there are other viable avenues for purchasing wheat, rice, and beans in large quantities, but I doubt they're as economical as what's provided through the investments of the Church under the direction of the prophet. If you have a Home Storage Center or dry-pack cannery in your area I highly recommend that you check it out. It's feasible to purchase an entire food storage without ever stepping into a Church cannery, but I think you'd be missing something.

It's actually a pretty good time

When I take friends and family to the dry-pack cannery for the first time I tell them it's like Lucille Ball meets Laverne and Shirley—an afternoon in aprons and hairnets like no other, and it works for men and women alike. My Primary boys especially loved the giant box stapler. Safety first. In conjunction with our lesson on Noah and the Ark I was trying to teach them the principle of listening to our modern-day prophets. Months later, I still get comments about our crazy fun party.

And the Spirit's there

Best of all, making the effort to just get myself to the cannery for work on my food storage feels surprisingly similar to serving in the temple. I experience the beauty of the Spirit in work that's bigger than myself. There's a definite feeling of strength and peacefulness inside the building that helps me provide for the temporal needs of my family.

I learned on the fly

The first time I went, I worried about not knowing what to do. Didn't need to. The cannery is staffed with volunteer service missionaries who quickly showed me the few easy steps to operating simple machinery. The hardest part is simply showing up with my checkbook. Walk-ins are welcome: I don't have to wait for the Relief Society clipboard to be passed around to schedule an appointment. If I need to purchase some canned dry goods I look for a day when I can steal an hour or two, make a phone call just to be sure the cannery's not too busy, and then go. I may choose to pop in for fifteen minutes or work for hours if I like.

Dry goods make great economic sense

During the early years, when our income was most limited, we purchased the cannery items that stored the longest and hoped they would spread the farthest. As struggling students and new parents, we quickly learned the economic reality that fifty dollars could buy a hundred pounds of plain white rice. That's four *months* worth of grain for one adult. Compared with how many small cans of soup we could buy at the grocery store for the same fifty dollars and how few meals we would gain from that purchase, we were sold on the basics.

So I'm not a bit ashamed about being a purist anymore. There are advantages to what the prophet suggested as "the basic foods that would be required to keep them alive if they did not have anything else." When food storage funds are limited, cans of wheat, rice, and beans—the boring staples—are the best long-term investment. These dry goods may be purchased one can at a time and should be usable for thirty years or longer.

The cannery may be uninteresting to "us" but it's completely remarkable to "them"

A couple of years ago I had the opportunity to take a Chinese woman to the dry-pack cannery on Welfare Square during her first visit to the United States. We had hosted her daughter through a University of Utah program that helps students learn about our culture. It was a year-long commitment and we were not supposed to do any proselytizing. Fine. As soon as our year was completed, we continued to invite our friend, Jean, to family dinners and activities. But now the gloves were off.

Free from the university program guidelines, we wanted to help Jean learn as much as we could share of the restored gospel. Eventually her mother, head of business and economic development for their city in China, came to Salt Lake City to visit her daughter after about six years of separation (and we think two years is a long time!) Considering that Jean's mother worked in city management, instead of a tour of the temple grounds I suggested that a trip to Welfare Square was the perfect way to introduce her to the Church.

Jean's mom asked questions like, "What happens if these facilities aren't profitable?" and "How can they run primarily through the work of volunteers?" I did my best to explain the systems in the Church and told her how grateful I was that my religion offered me the tools, assistance, and freedom needed to best provide for my family. You should have seen her face. You should have seen her with the hairnet, the apron, and canning her own #10 can of white rice.

At the bakery, both mother and daughter were offered a freshly baked loaf of bread. It was the total red carpet treatment. Employees, volunteers, and service missionaries were perfect in their efforts to be open and welcoming, and to share all that they could of this remarkable resource.

If you haven't visited one of the Church's dry pack canneries you should go, since it's inspiring to see what the prophet has made available to assist us in our food storage efforts.

Planning for "us": notice what you're already eating

In addition to the prophet's inspired direction on basic long-term storage items, each family is counseled to build an expanded supply of food normally consumed as part of their daily diet. This storage should uniquely suit their needs.

Despite my lack of cooking skills, I know better than anyone else what my own family likes to eat. That would be the food I'm preparing every day. And one day I realized that instead of constantly struggling to hunt down food storage meals we could stand to eat, the meals I normally cook—using foods my family already likes—could actually become part of our food storage plan.

Design a three-month supply from dishes your family likes to eat right now

Spaghetti was my first step on this leg of the journey, this one easy meal launching our design for food storage we could use on a daily basis. Using pasta from the cannery already stored in the basement, all I really needed to buy for our everyday storage were three jars of our favorite marinara sauce. Done. With that little effort I had the storage ingredients for one meal, to be served once a month, for three months. And look, a food storage dinner of spaghetti isn't even scary or weird!

It doesn't have to be complicated

If I were an amazing cook I would have of course bought all the *ingredients* to make my *own* homemade sauce. If that's your talent I seriously hope you'll go for it, and invite us to dinner. That's not me. Most of the time I'm working my head off, in and outside the home, and don't have time to make meals into recitals. I want recipes that are super quick, relatively healthy, and good enough to serve to company. Anything that can be made ahead of time is a fabulous bonus.

Just think of your favorite recipes and try to keep it simple
Following the spaghetti, I came up with one meal that used rice. Again, the design was easy. Who doesn't eat rice? But do we think of a meal with rice as *food storage*? It is. At one of my wedding showers, a zillion years ago, I was given a simplified recipe for Beef Bourguignon. It calls for sliced round steak, one can cream of mushroom soup, a package of Lipton onion soup mix, and a can of mushrooms. Sour cream is optional.

I already had the rice in #10 cans from the cannery so I bought 3 cans of soup, 3 envelopes of the dry soup mix, and 3 cans of mushrooms. Look at that! I had most of the ingredients for another dinner, to be served once a month, for three months. P.S.: When I find canned beef chunks on sale, I buy those as a substitute for the round steak. The fresh meat is better, but the canned wins in time and ease. This dish can be ready to eat in less time than it takes to have pizza delivered.

Make it quick
With those two dinners planned, I moved on to potatoes. While instant mashed potatoes aren't a grain or legume I felt they would be valuable considering our family, since we still need "kid food." My first mashed potato meal included mashed potatoes and barbequed chicken. For a three-month storage I only needed to buy three bottles of barbeque sauce. Ta-da!

Don't make yourself crazy

I wasn't worried that I still had to buy chicken breasts for this food storage meal. The goal isn't to end all trips to the grocery store, and we haven't been asked to build bomb shelters. Food storage can make sense for my life as it is *today* as well as having potential in the future. I'm not interested in living in fear of a catastrophe but prefer the example set by the pioneers. Those gals were tough and smart and capable: it's my sense that they understood the value of preparation and stored as much of their food as possible; then, as they were able, they added fresh items to their meals.

Make it fit your life

Beans were next. I actually began by purchasing cases of canned chili. Yikes. Not to be a total slacker, I served the chili with all the fresh fixings for taco salads. Today I have both dry beans—which I buy from the cannery—and cooked, canned beans from the grocery store. Depending on how much preparation time I have, I like having both options available. I also buy all the other dry ingredients needed in my favorite bean recipes for at least a three-month storage. Now I can make my own soups, enchiladas, and chili. I guess I've grown up some.

Be brave

Last of all, I decided to tackle wheat. When people tell those scary stories about what happens if you *eat wheat*… I try to stay calm. The *really* scary story is that our diets are in fact so messed up that we *can't* eat wheat. I don't think that wheat is the problem. My answer was to begin cooking one dish every week that used pre-cooked, whole wheat berries. That way my children could become used to wheat, both mentally and physically, and hopefully get to like it, which they easily did.

Be innovative

When we lived in Las Vegas a friend served a wonderful hot chicken salad at a baby shower. I use the same recipe in my food storage but substitute pre-cooked whole wheat berries for the Uncle Ben's wild rice originally called for. This recipe also uses fresh onion, celery, and green pepper, which I buy at the grocery store. The can of cream of chicken soup, mayonnaise, water chestnuts, and even the canned chicken all come from my storage. Yes, I really use canned chicken, and it's so wonderful my daughters prefer it to the diced chicken breasts I prepare myself. Maybe that tells you something about my cooking skills. Whatever.

Head in the game

I think the key to successfully designing highly eatable and useful food storage is to begin with five simple meals—just as I began with Spaghetti, Beef Borgie on Rice, BBQ Chicken with Mashed Potatoes, Chili Taco Salad, and Hot Chicken n' Wheat. When the ingredients for these five meals were purchased and stored in my cupboards, I started over. Continuing the same design process, selecting only one meal in each category (wheat, rice, beans, pasta, and potatoes), I kept expanding my storage, five recipes at a time. Eventually I had thirty family dinners planned, stocked in multiples of three, and stored for everyday use.

Planning meals is pretty boring stuff, so I chose to approach it as if I were packing the food for a trip to the cabin, beach, or Lake Powell. Faith in every footstep! Organizing the ingredients for five meals still felt like a stretch but in pretending I was preparing for a vacation I could muster the initiative to sit down at the table and get it done. Planning for a trip—something I already did several times each year—just felt easier than planning for some unknown pending doom.

I don't know a single person who prepares food for a vacation by just buying lots of miscellaneous ingredients, hoping to figure out the menus when they get there. I don't care how creative you are, no one does that for a vacation so why do we think that process would be smart in the event of an *emergency*? Sure we might be happy to just have something, or anything, to eat but worst case scenario, if things are in a state of chaos, *more than ever* I want to know *exactly* what I can do to provide for my family. For me, that kind of confidence takes practice and getting really comfortable with using my food storage.

Anyone can manage their own convenience store

Now we had food we liked and were happy to eat as part of our food storage. I felt like I was managing a mini-restaurant (okay maybe only a 7–11) but I knew exactly *why* I was buying these various ingredients for my food storage, I knew *how* I was going to use them, and I knew my family would probably *enjoy* the meals. I also knew that the money previously spent buying wheat, rice, beans, pasta, and potatoes at the cannery was not going to waste. My list of thirty dinners, designed specifically for my own family, was an organized system for incorporating food storage meals into our daily diet.

Can you see it?

With one daughter married and another out of state, we're now a family of five, so my current goal is to store ten cases of each of the recommended products that suit our preferences. Since we use the cannery for its convenience and economy, the total price for ten cases of wheat (as of this printing) is $142.20. Ten cases of rice costs $185.40. Over time, these relatively small amounts can be carefully factored into a family budget. Even if I can't afford to buy the more expensive everyday ingredients, for a very minimal investment my family will always have something to eat. Just add water.

There's no reason to lose it

Now I prefer to focus on *eating* the meals from my food storage rather than *rotating* my storage. Rotating sounds more like a dance...with cans...but I clearly understand what eating is about. Usually I don't even take time to write a date on my grocery store cans. No! It's true. I really don't need to. I'm cooking with those ingredients all the time, so they're naturally getting used up and replenished.

Check it off the list

Just so you know, I've made peace with my food storage but there's no way I cook every night of the week. There has to be room for leftovers, dollar menus, cold cereal, and the occasional restaurant. I want food storage to bless my life, not run it. Staying on track, regardless of the normal periodic gaps, is easy because I maintain a running list of my everyday food storage meals. I select at least one dish from each category (wheat, rice, beans, pasta, potatoes) and quickly note the few fresh ingredients I'll need from the grocery store. As I prepare and serve those dinners during the next week or so, I simply check them off the list. I'm keeping myself accountable.

Meals don't have to be prepared in a predetermined order. Flexibility is vital. My schedule and the family's needs are constantly changing. By selecting five to seven meals at a time, the meats and produce I buy at the grocery store stay fresh as I consistently work my way through both our long-term and three-month food supply. I consider what sounds good and fits in with life during the next week, but the big idea is that I've brained it out once and stick with the plan. I've learned that being a more efficient manager of our food needs has literally freed up both time and energy in my life.

Knowing what to serve for dinner has always been a pain
And speaking of my life, let me share just some of the ways food storage has made my job—the job I do at home providing the meals—so much easier. Before using this system I probably wasted years dragging myself through cookbooks, foraging through cupboards, trying to figure out, *again*, what I could make with the cans of tuna, box of Jell-O, and jar of mustard I had on hand. "What are we having for dinner?" "I DON'T KNOW! STOP PRESSURING ME!" Okay, a tiny dramatization but you get the point.

*Something's wrong if I find myself doing
the ridiculous three and four times a week*

Besides all that goes into planning meals and coming
up with a weekly list for the grocery store, I have to
tell you that I hate the actual shopping part. I hate the
whole process of making the list, dragging myself to the
store, parking, hunting up and down the aisles, searching for random ingredients, and then waiting in line to
hand over money. Having an expanded meal plan has
meant that all the nasty planning is already done. I can
go months, and months, and months between my 10-K
shopping runs where I buy as much as I can afford in
groups of three for my three-month storage supply.

*It's a lot easier to focus, design a
viable program, and then stick to it*

Designing meals that make sense for my own family
means that I become really comfortable preparing those
dishes again and again. A shift in lifestyle takes commitment and real work, and I believe that understanding how to prepare and use food storage is a very smart
shift. It all comes down to preparation, and there's so
much more to it than just shopping and buying random
piles of extra food.

Now I'm not constantly re-designing or re-engineer-
ing our menus. I would never tolerate that pattern in my
business. The repetitive tasks, the mundane details that
have to be handled day after day get figured out ahead of
time. If I handled my jobs the way I previously handled
cooking for my family I'd be out of business. That's why
we have QuickBooks, file cabinets, and planners. Systems help us to be more efficient and effective: being

organized in the management of my kitchen is a basic concept, but it took a long time for me to learn how to use some of my professional skills in my own home.

Food storage starts out costing money and then it saves, and saves, and saves

Today, when I make a run to the store it is so much quicker. I simply grab the dairy items, a little meat, and then a few fresh fruits and vegetables. In addition to the time-saving aspects, our food storage plan absolutely saves money. We buy in bulk and enjoy the maximum value of sale prices. And we go to the store less often. Fewer shopping trips help liberate us from impulse purchases. We also save because we spend less money on fast food. I don't have to grab something on the way home since I already have easy options, ready to go—at home. It's like the prophet was inspired!

I can grow my system as much as I like

When I come across a new recipe that we enjoy and want to incorporate into our storage, it can easily be added. We don't have to eat the exact same thing every single month. But simplicity is also important: I like a nice variety of meals, but when it comes to organization, less is always more. I'm currently working from an expanded list of forty dinners. Considering the natural gaps in cooking, forty dinners means we repeat the same meals only about every two months. And if my family needs more variety than that, they need therapy.

The rest of the story

These food storage dishes rarely make up an entire dinner. I always try to add fresh fruits, steamed vegetables, and great salads. It's primarily the main course ingredients that are stored. Those items are the priority. I also buy a variety of canned fruits and vegetables, dried fruits, some nuts, and our favorite salad dressings for our everyday food storage. This way I have even more of the basics on hand for the things we normally eat for dinner. I don't break the side dishes down into specific recipes; I just buy things we commonly use, by the case, when I find them on sale.

And of course, dinner isn't all we eat. Just like you, my family likes three meals a day, so I've made some preparations for breakfast and lunch as well. Breakfast can be oats or wheat cereal. (Add ten cases of regular oats to my dry-pack shopping list. It should cost about $120, and that's a big step towards breakfast for five adults for a year.) Oats and wheat are at the top of every list of "super foods" I've seen, and eating them for breakfast is the easiest way to add them to our diets. Lunch can be tuna fish or peanut butter and jam. I can't get caught up in micro-management of every last detail. I believe food storage is meant to make our lives *easier*, and *less* complicated.

Planning for "them": consider how you can share with the neighbors

So many of us have been blessed to know there is a prophet of God on the earth and yet there are so many more who have not, for whatever reason, as yet had this great light in their lives. I don't know how all that was originally decided but I feel a deep responsibility for the knowledge I have been given. While I don't secretly hope for some personal or national catastrophe, I do hope that in the event of an emergency I would be able to help people around me. And I hope that if I could offer them food they might want to know how I knew to store it ahead of time and thus hear my testimony of a living prophet. Ultimately, I believe food storage is about love—love for the Savior, love for the prophet, love for our families, and love for our neighbors. Perhaps there is an eternal responsibility in sharing food and sharing the gospel.

When planning food storage it's important to consider how we can meet the needs of those beyond our family. Today the neighbors might think it's a little nuts to keep all that extra food on hand, and that's okay. If something happens tomorrow, food storage will make perfect sense to everyone.

They worry about their children, too

We lived in New York during the Y2K panic. One day, a friend of our twins was commenting on these "freaks" she had seen on the evening news: "They have all these cans and cans of food, just stacked up in their basement." Mouth open, I flipped my head to catch our daughters' faces light up with joy as they burst out laughing. Unable to form sentences, they grabbed this friend by the hand and dragged her down the stairs into our tiny hole of a basement. I can't even repeat what she kept yelling as she took in our shelves full of food. We all laughed so hard—didn't even try to explain to her the rationale.

For me the funniest part happened a few weeks later, with a group of parents at a school function. Our twins

were the only members of the Church at their middle
school, so we already stood out. The news of our food storage had spread and there was some lighthearted joking
as parents gave us a friendly, hard time. Before we left
for home, a couple of them quietly and "all joking aside"
asked if they could come to our house if things turned
freaky with the start of the New Year in 2000. Of course
we answered, "Yes!" And no, we didn't need to share our
food storage at that time, but maybe what mattered was
that we were prepared and willing to.

Preparation means I have more to offer

We know that God loves "them," or our neighbors, just as
much as he loves "us." A neighbor's food storage may be
precisely how the Lord is planning to help and provide for
"them"! One day in June, Joe, who lives next door, called
to tell me that another neighbor had just come home from
the hospital and was recovering from hip surgery. The
neighbor doesn't happen to be a member of our Church
but could of course benefit from a home-cooked meal.
Using my food storage I was able to throw together something hot and tasty within an hour, and I didn't have to
go to the store. That day, with everything else I had going,
having the ingredients on hand made all the difference. I
was better able to help because I was better prepared.

And what if I can't get to the store?

On a larger scale, say there was a major crisis and more
of my neighbors needed food. If anything like that happens, everyone knows the stores will be empty in an hour,
so I probably won't be able to get fresh ingredients. How
do I deal with my everyday food storage recipes now?

In an emergency, I plan to boil a pile of wheat and mix it with piles of rice. Synergy. Then I select one of my food storage recipes, "Hot Chicken and Wheat Casserole," but I forget about the fresh onion, celery and green pepper. Instead, I simply combine and heat one can of chicken chunks in its broth, a little mayo, one can of cream of chicken soup, and maybe a can of water. Everyone at this party gets a healthy serving of wheat and rice—"storage pilaf"—with one small spoonful of chicken gravy on top. Voilá!

That emergency meal would be familiar to my own family, hopefully tasty to others, and most importantly do the job and serve many people. As I select meals for my three-month storage, I evaluate recipes and consider up-front how useful or adaptable they would be if I couldn't buy all the fresh ingredients called for. I plan to do all I can to make my food stretch as far as possible and then pray like crazy that the Lord will do the rest.

Notice the very first paragraph in the "All Is Safely Gathered In: Family Home Storage" pamphlet. It opens with this message in which everyone is counseled to get prepared and assist in the care of others:

Our Heavenly Father created this beautiful earth, with all its abundance, for our benefit and use. His purpose is to provide for our needs as we walk in faith and obedience. He has lovingly commanded us to "prepare every needful thing" (see D&C 109:8) so that, should adversity come, we may care for ourselves and our neighbors and support bishops as they care for others. (Intellectual Reserve, Inc., 2007)

If I don't already know that *my* food storage is really the *Lord's* food storage, it's only because I haven't been listening. Clearly, our leaders have always meant for food storage to be shared. The law of consecration is taught in the temple, where promises are made every day. Bishops calling for members' food storage may have been soft-peddled in official pronouncements, but we know the Lord trusts his saving work to be accomplished through the efforts of regular people willing to sacrifice for each other.

I'm thinking about these possibilities while I'm designing my food storage meals. Beyond dishes my family enjoys, I'm selecting meals that transition well and praying for the heart and sense to freely share if that time comes. It takes an eternal perspective to make room for the needs of others, and I don't have a perfect record in that area. Visiting teaching has been neglected on more than one occasion, and that's when I've only had to sacrifice fifteen minutes of my *time*. What's going to happen when I need to sacrifice food? Truly the Lord's grace will be essential in order to share food when concerned that my own children don't go hungry. Storing faith along with the food is what the spirit and letter of the law look like, to me.

Having an eternal perspective involves praying for charity
Prayer has a role to play in preparedness. Hearts have to get ready, too. I admit that at the deepest level I'm mostly concerned for my own children, but I want to become the person who completely and perfectly cares about all children. The scriptures counsel us to seek after the best gifts and to be specific in our prayers. My daily prayers

include a plea to discern the needs of others: I ask for charity because that's a gift from God, and I ask Heavenly Father to give me a portion of the love both He and the Savior feel for all people on the earth.

And what about the prayers we offer over our food? Ours can become pretty Zoramite-ish as we rattle off the same memorized phrase three times a day:

> Dear Heavenly Father. We thank thee for the food we have to eat and ask thee to bless it to nourish and strengthen our bodies and do us the good we need.

I bet my prayer will be very different in an emergency:

> Thank you, Father, for our lives. Thank you for the food we have to share. Thank you for a living prophet who lovingly warned us for years and years that we should prepare and store this food.
>
> We humbly ask you, Father, that you will bless and stretch our food. Please bless this food, as we share it, that it may truly nourish and strengthen as many as possible of your children.
>
> Please help us to be filled and generously reach out to serve one another. Please help us to be unselfish. Please bless us with the comfort of the Holy Ghost and the courage to follow the guidance we receive. Please be with all of us and help us to work hard to be good brothers and sisters to each other, that we may become fit to return to thee.

I believe that if we've honestly done what we can to prepare, we can ask our Heavenly Father, in faith, to bless and expand our food storage so that it may be nourishing, satisfying, and sustaining as we strive to help one another during a crisis.

It's not just about providing for our own families

Praying about my food storage has brought specific answers about how I can prepare to help others. For example, promptings from the Holy Ghost have made me reconsider my position on powdered milk (the last frontier). For my own family I've mostly disregarded the idea of powdered milk. Besides purchasing a few token boxes, I knew my daughters wouldn't drink it and rationalized that the counsel didn't apply to me. I would buy everything suggested, but not that. My daughters would insist powdered milk is precisely what the prophet was referring to when he said, "Such food may not be tasty, but it will be nourishing if it has to be used."

A friend once gave me a book of recipes using powdered milk. As I flipped through it I noticed one recipe that used a token amount of powdered milk in a delicate and very complicated cream sauce to be served over...swallows. Now, it's a great book and the perfect gift for me, but if I'm in an emergency, and before I make a cream sauce to be served over...swallows, I'm just going to drink the powdered milk, by the glass, straight up. Please! It's truly not that bad.

Lately I've started to think a bit more about "them" and wondered if I couldn't do something to help prepare for 'their' needs. How hard would it be for me to store some milk that could be shared? Most recipes calling for milk will have fine results with substitution of powdered milk. Maybe I could buy powdered milk, start using it in my cooking so that it doesn't become a waste of money, and have it on hand as the prophet has asked. Maybe I could be the one to have it stored and be a help to someone with little children—babies—who really need some form of milk. Maybe my food storage doesn't have to be just about "us." Maybe I can also store what I would like to share.

My food storage can help others weather a catastrophe

Recently I completed a wonderful religion course taught at the BYU Salt Lake Center. From our text, I found an inspiring bit of history illustrating how women of the Church responded to the needs of people in Europe following the Second World War.

The book describes how American Relief Society sisters were "uniquely prepared to help provide food for the starving peoples of war-torn Europe" because they had already been storing wheat for years. When the emergency hit and people across the ocean were desperate for food, these faithful women sold two hundred thousand bushels of wheat to the United States government, which shipped the grain overseas. The Relief Society then used the proceeds to help maintain a special "wheat fund": "continuing to buy wheat, they made sure that a basic food source was available for future charitable purposes" (*Church History In The Fullness Of Times*, Church Educational System, 2000, p. 492).

We're living in a time of war right now, and natural disasters happen more and more frequently. I've heard a representative from the Red Cross state that they depend greatly on the food supplies that come from the members of The Church of Jesus Christ of Latter-day Saints. We pray for the safety of our soldiers and we likewise pray for the safety of all our brothers and sisters—children of our Heavenly Father affected by conflicts all over the world. Having an available food storage is another way we can lend our individual support. We may or may not be called upon to share our supplies, but our preparations are tangible evidence of a desire to help.

Speaking of the Red Cross

Very late one night in New York, we received a phone call from our bishop. He told us the stake president requested that members immediately assist at the site of a huge fire, which had burned a substantial portion of an apartment complex. The Garth Road apartments in Scarsdale were just minutes from our home, and as soon as we arrived we began preparing food for the residents while collecting necessary information for the Red Cross.

Moments after our arrival, an official asked for help locating people trapped in their apartments, without electricity, and I found myself staring at the double door entry of one of the buildings. The wide marble entry stairs looked so strange with water pouring over them like a fountain. A fireman handed me a flashlight, assured me it was safe to go inside, and asked that I knock on doors and assist people as they exited the building. Carefully I stepped down the black hallway in water about an inch deep and called "hello," again and again, when one door opened up and a young couple peered into my light.

While I explained details of the fire and the evacuation, they quickly tried to collect some clothes and business papers. I offered to carry as much as I could, and the three of us very carefully splashed our way down the long corridor and out into the flashing lights and noise on the street. Standing on the sidewalk, the couple informed me that it was their one-year wedding anniversary and they had just returned from dinner when their electricity was shut off. I felt so sorry for them and offered our home as a place they could spend the night; they politely insisted that friends would take them in, and we soon parted ways.

The part I wish I could express about that night is how it felt to really be needed by these strangers and then to help and serve them. I didn't do very much but I did all that I could. All I had to share was a little light, but because of that light it felt like we were instantly more than best friends. It felt like we were brothers and sisters, and I deeply wanted to help with their burden. Being able to share is really an opportunity that brings both love and joy.

Don't even go there

Have you ever heard someone say, "I don't need to worry about food storage because I've got ammunition"? I know it's a joke. It's a really old joke and never very funny. Talk about acquiring or defending our food storage with guns is crazy talk and it scares me. When it comes from a faithful member of the Church I wonder why we're even bothering with attending Gospel Doctrine class. We can share our food. The situations in the scriptures that deal with food, or the lack of food, are consistently examples where the faithful people who were serious about following the Savior's example shared their food. I've had to ask myself what good it will ultimately do to sustain the lives of my family and myself if I have turned my back on others in need. I don't believe there will be eternal rewards for that choice. That's why I pray about my food storage silently at the grocery store and on my knees at home. I'm going to need more strength and more charity, and I'm seeking it early.

As I've designed my own food storage, then, my goal has been to plan to share it with others, to consider the needs of my family, and to follow the counsel of the prophet. The next chapter will walk you through that process of designing your own food storage and, in a practical way, help you to implement your own food storage plan.

START NOW

Implementing Your Food Storage Plan

Organizing your own food storage is easier than you think. It is simply a step-by-step process that incorporates actions and activities you're already doing.

Designing a comprehensive food storage plan

As I'm a visual learner, I've created a chart illustrating my concept of how a long-term supply, a three-month supply, and fresh ingredients combine to create a comprehensive food storage that can be used on a daily basis:

Eating Your Food Storage

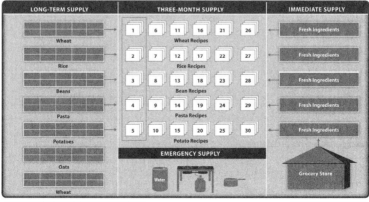

Long-term supply = Dry goods that store 30 years or longer
Three-month supply = Everyday dried, canned, and bottled ingredients
Immediate supply = Fresh dairy, meat, and produce
Emergency supply = Water and basic cooking supplies

Long-term food storage

A long-term storage supply is comprised of basics, most of which can be stored for thirty years or longer. The Church recommends that for a year supply a family of five store 1,500 pounds of a variety of grains, in whatever combination they prefer, plus an additional 300 pounds specifically in dried beans. I selected the six basic items my family is most comfortable using and then doubled up on our storage of wheat, because it lasts the longest, is inexpensive, and guarantees that some form of bread or cereal will always be available. These seven groups of staples, each with ten cases, slightly exceed the 1,800 pounds total weight suggested on the Church's "Provident Living" website (www.providentliving.org). This particular structure gives a long-term food storage a nice variety of basic products and is simple to manage and inventory.

For your family, if you multiply the number of people by two cases in each of these seven groupings, you will reach the weight goal recommended by the Church. For example, the chart illustrates that a family of five, multiplied by two cases, needs ten cases each in seven food groups (5 x 2=10). Using this same format, a family of three (3 x 2=6) needs six cases of wheat, six cases of rice, six cases of beans, six cases of pasta, six cases of potatoes, six cases of oats, and another six cases of wheat to have the basics completed for a year supply of long-term food storage. It's very simple.

Three-month supply

The "Three-Month Supply" section represents thirty different meals that incorporate a long-term storage item. I recommend designing these dishes five at a time, building up gradually to the goal of 30—one month of meals, to be repeated three times—so that the planning or financing doesn't become overwhelming. Select meals that suit your family's preferences, and then purchase the short-term storage ingredients in groups of three for a three-month supply of everyday meals.

Immediate supply

The "Immediate Supply" section includes fresh or frozen ingredients purchased on a frequent basis at the grocery store. The 30 meals I've designed for my family use our long-term storage, our three-month storage, and fresh ingredients from the supermarket. It's actually the fresh ingredients that make the thirty meals viable on a daily basis. The fresh ingredients help our food storage dinners to be *normal*. Remember that the prophet has never asked anyone to stop going to the grocery store. All we've ever been counseled is to gradually build up our own supply of foods that *can be* stored, so that we, our children, and the people around us, are not completely *dependant* on the grocery store.

If you use a sheet of paper to cover up the portion of the diagram containing the grocery store and fresh ingredients, you can see that in the event of an emergency, food storage is still a workable resource with just the long-term and three-month supply. Recipes may not be prepared exactly as originally planned, but there is a wide variety of familiar ingredient options that could greatly enhance the basic grains and legumes, thus helping to comfort and sustain during a time of need. With water and some form of emergency cooking source, both of which have been around from the dawn of time, a family has what they need to assist their own and serve others if normal supplies are interrupted. (For specific guidelines on water storage, see the Church's Provident Living website: www.providentliving.org.)

So get going—you can do this

In the following sections—the actual instruction part of this food storage book—I've tried to offer precise information regarding the tools I've used to organize my efforts in provident living. With as few as six recipes in each of five long-term storage categories, you can systematically develop a personal repertoire of thirty meals. Most of us probably already use about that many recipes on a regular basis anyway. Defining your top thirty meals means you always have a month's worth of options, collected, organized, and neatly compiled. Finally, you can quit wondering what your family should eat for dinner.

The seven steps below will walk you through the process of creating a food storage you can actually eat. You'll also find some simple work-pages that you can complete to help you get organized. Some of the steps may seem unnecessary for your needs, but they've helped one slacker stay on track, and that kind of success speaks for itself.

Step 1: Shop for your long-term storage

While considering your current financial situation, determine what steps you can take, and get moving toward a one-year supply of the long-term storage items used in your recipes. A reasonable goal might be to meet at the cannery, as a family after work, once a month for just an hour.

A. Purchase one case (six #10 cans) each of wheat, rice, beans, pasta, potatoes, and oats, with either a second case of wheat or rice, totaling seven cases. Seven is easy to remember because there are seven days in the week and that's about how often most of us like to eat. If you choose to use the Home Storage Centers, the canning process should take you an hour and cost around $120 (think of seven trips to McDonalds). Store in your kitchen one opened can from each case of wheat, rice, beans, pasta, potatoes, and oats. Long-term storage items must be easily accessible or they'll be forgotten.

B. Resolve to purchase one additional long-term storage item each month, continuing to build your long-term supply while you're using your current food storage. Set realistic goals for both your staples and extra supplies, and schedule specific dates for food storage purchases. You can use the following equation to figure how many cases of each food category you will need, based on your family size:

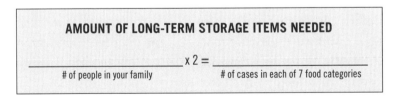

AMOUNT OF LONG-TERM STORAGE ITEMS NEEDED

$$\frac{}{\text{\# of people in your family}} \times 2 = \frac{}{\text{\# of cases in each of 7 food categories}}$$

Counsel from the First Presidency in the "All Is Safely Gathered In" pamphlet gently states, "We encourage you to store as much as circumstances allow" (Intellectual Reserve, Inc., 2007). Each one of us has to determine what's appropriate for our needs. Food storage is

an individual choice, a personal effort, with an ever expanding potential to bless. Dedicating one hour at the
cannery and less than about $120 per month to provi-
dent living preparations, a family of five could complete
their one-year supply of the basics in ten months. After
that they would still have a couple of months left in the
year to focus on additional items like powdered milk,
sugar, oil, and salt.

Sometimes I hear people question whether we can
really eat any of the hard-core food storage. It seems to
be their opinion that the basic staples are too old-school,
and cooking with them has become a lost art and obso-
lete in our day. In my head I'm thinking, "just add water":
that's how all of the basic staples cook—with *water*. I see
it as a great advantage that the foods that are the least
expensive, and can be stored for the longest amount of
time, all cook with water; it couldn't be more basic. The
fact is, if you can boil water, you can cook and eat your
long-term food storage.

Step 2: Get organized by building your own food storage binder

Before planning meals using your long-term food storage, build an everyday food storage binder to organize
the basic categories for your family's meals. The binder
can be your food storage directory.

A.   Get yourself a one-inch, three-ring binder, five sheets
      of colored card stock, and a package of fifty plastic
      sheet protectors.

B. Select five basic food categories for your family's
    meals. Divide meals into categories that make sense
    for the way your family eats. I've based my catego-
    ries on the wheat, rice, beans, pasta, and potatoes
    that are offered by the Church. Label each of your
    five sheets of card stock with one of your chosen cat-
    egories, and then place it inside a plastic sheet pro-
    tector and into the binder.

Food storage binder organized by long-term categories

Step 3: Choose one recipe for
each of five long-term storage categories

A. Select everyday recipes your family enjoys as the
    foundation of a three-month food storage. Try think-
    ing of one at a time, using the Internet and asking
    friends for their suggestions for *normal*, tasty meals.
    If you don't have a ready recipe for one of the cat-
    egories, you might try one of those listed in Chapter
    4. Selecting recipes suitable for food storage begins
    with two basic criteria:

1. The dish should be quick to prepare. Simplifying life matters a lot to me right now and I have a hunch that simplifying will matter a lot during a crisis. Sure, I still use recipes for special occasions that require much more effort—and tarps and total silence—to prepare, but those are not part of my food storage plan. We love fancy recipes, but they're not the standard. I'm always on the lookout for quick, tasty dishes. Whenever we eat dinner at a relative or friend's house and they serve something yummy and then tell me "Oh, it's so easy to prepare," I'm planning on asking for that recipe. Almost every one of my food storage recipes can be traced back to a dish that's been shared. Collaborating with friends is one of the best ways to find easy recipes.

2. The recipe should call for both dried and canned goods. There isn't any rule of thumb for the number of storage ingredients. Like I've mentioned before, I select *normal* recipes that incorporate combinations of fresh vegetables, fresh meats, dairy items, and a few storage ingredients. Sometimes I modify recipes, smuggling in whole-wheat berries or substituting canned meat if I think I can get away with it. Most importantly, if a recipe has been added to my list of 30 meals I've already considered how only the short-term and long-term storage ingredients could be combined, in the event of an emergency, to create an *enhanced* dish of rice, wheat, pasta, etc.

I'm planning that my three-month storage will lend variety and make my long-term storage more interesting, especially if we have to live on it for an extended time. There will likely be changes in how each recipe is prepared, but I'm betting that the combination of flavors will still seem familiar to my family, and that they will enjoy it more than plain old anything (remember the Children of Israel eating manna in the wilderness? Those people had food; what they hated was not having any variety).

Even though oats are one of the long-term staples, I don't include recipes for oats in my three-month storage plan meals since I don't serve oats for dinner. I eat them almost every morning for breakfast and I don't need a recipe for that. Periodically I use them in apple crisp and cookies. I don't need to plan that either. Dessert happens.

To begin your own food storage plan, choose a recipe for each of the long-term categories you selected in Step 2 (the defaults here are wheat, rice, beans, pasta and potatoes) and write the names of the recipes below:

FIRST 5 STORAGE RECIPES	
Storage category	**Recipe name**
Wheat	_____
Rice	_____
Beans	_____
Pasta	_____
Potatoes	_____

B. Place each of your five everyday recipes inside a plastic sheet protector and then into the food storage binder, behind the appropriate meal category divider. They could be ratty old originals or photocopies made from a vast collection of recipe books, but placement in a designated binder gives clarity and focus to your food storage. Having your recipes together will help you to stay organized and help you save time as you actively use your food storage.

Step 4: Create a three-month storage inventory and shopping list

Everyday recipes transition to three-month food storage recipes simply by purchasing and stocking your shelves in larger quantities. Think five meals multiplied by three months: design in groups of five and shop in groups of three. Rather than purchasing cases of items your family may use periodically, you will be buying ingredients for meals you are regularly eating. Creating a three-month storage inventory and shopping list right from the start helps organize and maintain the supply. It also can help you save money as you watch for case-goods sales for items you know you will be using.

A. Take five sheets of paper and write the meal category (e.g. wheat, rice, beans, etc.) at the top of each one.

B. Divide each sheet into four vertical columns with headings for each of the following:
    1. Recipe
    2. Fresh ingredients
    3. Storage ingredients
    4. Three-month quantity

C. Select a recipe to begin with, one of the five meals you wrote down in Step 3 above.

D. Using your chosen recipe, record the following information in each of the four columns:
    1. Recipe: name of the recipe
    2. Fresh ingredients: each fresh ingredient required by the recipe
    3. Storage ingredients: each "storable" ingredient for the recipe
    4. Three-month quantity: each "storable" ingredient x 3

One entry might look like the following:

Three-month Storage Inventory and Shopping List			
Wheat			
Recipe	**Fresh Ingredients**	**Storage Ingredients**	**3-Month Quantity**
Hot Chicken and	½ c. butter	2 cans chicken chunks (12.5 oz.)	6 cans
Wheat Casserole	1 c. celery	1 can cream of chicken soup (10¾ oz.)	3 cans
	1 c. onion	¾ c. mayonnaise	18 oz.
	1 c. g-pepper	1 can sliced water chestnuts (6 oz.)	3 cans
	2 T. butter	1 c. bread crumbs	24 oz.

As you write down the three-month quantity of each ingredient, be sure to eliminate any guesswork while at the grocery store by noting

a.   the *number* of cans needed for a single recipe

b.   the exact size cans (8 oz. 12 oz., 10¾ oz.) called for

You can begin this step by filling in the first entry of your three-month inventory and shopping list, using one of the long-term storage categories and recipes you listed above in Step 3:

Three-month Storage Inventory and Shopping List			
Long-term Storage Category:			
Recipe	Fresh Ingredients	Storage Ingredients	3-Month Quantity

You'll notice that in the example above, as well as in the three-month storage inventory and shopping list offered in the "Resources" section that I don't ever list the amount of long-term storage ingredients—the wheat, rice, beans, pasta, and potatoes—required for recipes. Those items are part of a much larger storage, one that is always purchased in bulk amounts and supplies a much longer time period than just three months.

Once you've filled out the recipe information, you can transfer it to another document that you will use regularly. For instance, you might write it down neatly on a separate sheet of paper, or type it into a word-processing or spreadsheet program on your computer, or input it into an electronic planner. The type of document you create doesn't matter; what does matter is that you choose a method that works with your current organizational strategies so that it will be easy for you to use. See page 157 for an example of a three-month storage inventory and shopping list in the form of a spreadsheet, using the recipes from Chapter 4.

Step 5: Shop for your three-month storage

A. Go to the grocery store, with your three-month storage inventory and shopping list in hand, and purchase items in the "Three-Month Quantity" column, as your finances allow.

B. Shelve your items at home: place the cans and jars in your kitchen cupboards, rather than in a less accessible location, grouping items on the shelf by category and recipe (see photo). That way, as you move through your supply, ingredients needed for a specific recipe are together in one spot. As you run low on ingredients, simply compare what can be counted on the shelf against what's noted on the inventory list. This will tell you exactly which items need to be replenished and what quantities you should purchase.

C. Store the relevant portion of the inventory and shopping list with the food purchased for that category (wheat, rice, beans, pasta, and potatoes). You might tape it to the inside of the cupboard, or attach it to shelving with string: having each list right where the food is stored will make it easier to count the supply and restock items as they become depleted.

Shelving organization of food storage meals

D. Store a second copy of the inventory and shopping list in the back of your food storage binder as a back-up. You may also keep a copy in a daily planner.

Step 6: Expand the number of meals in each category

With the first set of five everyday food storage meals planned, purchased, and stocked, continue designing your own food storage meals so that there is a reasonable amount of variety in your family's daily diet. A recommended goal is thirty meals. That means collecting only six recipes for each of the five categories, and simply repeating the steps you've already completed.

A. Select one new recipe for each of the five food categories.

B. Repeat Step 3, placing each recipe inside a plastic sheet protector and then into the binder behind the appropriate divider.

C. Repeat Step 4, adding this new series of five recipes to the relevant categories of the inventory and shopping list you've already started.

D. Repeat Step 5, shopping 5 x 3 (five recipes' ingredients times quantities of three) and then organizing your storage ingredients by category and recipe in your home.

Expand your everyday food storage, five recipes at a time, until you're prepared with six meals in each of your five categories. This process might take you three days or three months to complete: simply work at a pace that you're comfortable with.

A. As you work through the process of building your list of recipes from five to thirty, you can record your chosen meals on the following chart.

30 Storage Recipes	
Storage Category	**Recipe Name**
Wheat	1. _____
	2. _____
	3. _____
	4. _____
	5. _____
	6. _____

30 Storage Recipes (continued)

Storage Category	Recipe Name
Rice	1. _____
	2. _____
	3. _____
	4. _____
	5. _____
	6. _____
Beans	1. _____
	2. _____
	3. _____
	4. _____
	5. _____
	6. _____
Pasta	1. _____
	2. _____
	3. _____
	4. _____
	5. _____
	6. _____
Potatoes	1. _____
	2. _____
	3. _____
	4. _____
	5. _____
	6. _____

Once you've got thirty clearly defined dinner options, with the majority of ingredients purchased and stocked in your home for a three-month supply, you've reached food storage Nirvana!

Step 7: Use a planner list to track your progress eating your storage meals

Resist the temptation to just store your food storage. In doing that you limit some of the practical, day-to-day blessings of preparedness and you risk wasting your investment. All food storage—especially a three-month supply—must be used and replenished in order to stay fresh. Eating your food storage is just as important as buying it.

A.   List the names of all thirty recipes in your day planner (creating a mini-index by copying the list you created above), and make a plan for eating those meals. You might decide to use your food storage recipes every day of the week or only a few times each week. What matters is that you set a reasonable goal and then consistently move toward that goal.

B.   Choose the meals you'll serve over the chosen time period, note the fresh ingredients required, and then complete your speed-shopping trip. This is when you'll purchase the fresh meat, dairy, and produce items that your recipes call for.

C.   Check meals off your list each time you prepare them. When you have checked off a recipe three times, it is time to shop to replenish your three-month supplies.

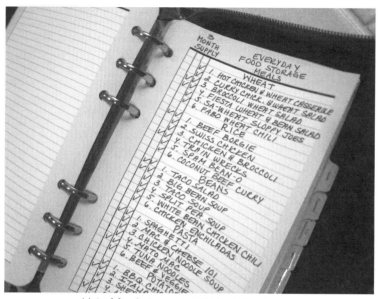

List of food storage meals in daily planner

As the weeks go by, this handy reference list shows at a glance which meals have or haven't been served. It keeps you from getting lost in the process. Check marks illustrate progress. Serving thirty meals, three times each, may take as long as six months to accomplish. That's not a bad thing. A planner list keeps everything on track.

4

TRY THIS

Easy Food Storage Recipes

Oh no—here we go with the whole cookbook thing

I've reluctantly included my recipes only as ideas or examples of what you can do with the very basic food storage items. These recipes may or may not fit your cultural background, dietary needs, or personal preferences, but I hope they will give you some ideas as you devise recipes that work best for you and yours. Perhaps sharing what I cook will help you see the "hard-core" food storage items in new ways.

But it's really not about perfect recipes

Recently, I helped with a dinner for all the sisters in our Stake, held prior to the General Relief Society Meeting broadcast. It was the mother of all potlucks, with every dish prepared from ingredients that could be part of a three-month and longer-term storage. Each ward was assigned a specific table where they offered six different dishes highlighting wheat, rice, beans, pasta, or potatoes. They also set up displays of the storage ingredients used in each dish so that sisters could sample the dishes and sneak a peek at the ingredients incorporated into everyday storage.

Those of us in the Stake Relief Society presidency were relieved that there was enough food, but I was also pleased at many sisters' genuine surprise that the food tasted so good. The only complaint I heard, and heard, and continue to hear, is that we didn't offer the recipes. That would be mostly my fault. I just couldn't stand to burn the time or money collecting all thirty recipes from five different wards and then shovel out more food storage handouts, when we know they too often land in the trash. I voted that all of us can stop fretting about finding the perfect food storage recipes and instead just become more aware of the recipes and ingredients we already use. All we need to do is buy what we normally buy, but in slightly greater quantities.

It's about what your family likes to eat

People can design food storage that's perfect for their own family if they'll just look creatively at what they already like to eat. A customized, personal food storage is so much easier to use and enjoy than a one-size-fits-all system. There's no way we're all going to eat the same things, and I don't want to try to tell anyone what they should buy. There's enough of that fun in my job as a designer. Anyone is welcome to my recipes, but it's of greater value to design most of your individual thirty-day meal plan. Just remember the two criteria for selecting meals for your own plan: they should be easy to prepare, and they should include both long-term (staples such as rice or beans) and three-month (such as canned goods) storage ingredients.

Begin your own list of thirty meals by reviewing the daily recipes you prepare for your family and considering how those same recipes could become a viable part of your own three-month storage; then you might fill in any gaps with the recipes below. In making the effort, you'll be more committed to what should be purchased, know you're prepared with food your family enjoys, and be better able to follow a reasonable, practical system for using it. (And, I'm hoping to encourage you all along the way.) So! Here we go.

First, let's chat about wheat

The recipes below purposely begin with wheat. I know there are a lot of myths about storing and eating wheat. Some people suppose that their bodies can't handle wheat. If you've been diagnosed with Celiac Disease, then it's true for you. The rest of us should remember that wheat is one of the least expensive grains, stores the longest, and is an excellent source of protein. It contributes dietary fiber, B vitamins, vitamin E, calcium, phosphorous, and iron.

But I'm just betting most people don't have a clue about how to cook with wheat and could use a few hints. I was that way. At least twenty years ago I made the big decision to go ahead and purchase wheat as part of my long-term food storage. I didn't know how to use it, and I couldn't make bread if you held a gun to my head, but for less than $200 for *seven* people, I chose to invest in the prophets' recommendations. I figured I had at least thirty years to learn what to do with it.

Time passed and I eventually learned how to make bread and cook with wheat. (I actually signed up for a class.) Now my daughters don't see whole-wheat berries as some sort of weird purist food. They think it's just another variety of grain as popular as brown rice, wild rice, or barley. Anything sold at Whole Foods has to be cool and really tasty, doesn't it? We haven't always felt cool.

One summer, we went to a great neighborhood picnic where everyone brought their own meat and a side dish to share. I decided to make my favorite Fiesta Wheat and Bean Salad and serve it with chips as a fresh salsa. Then, only because I had everything on hand and didn't want to go to the store, I chose to make Sa-Wheat Sloppy Joes as our family's main course. It was risky but I did it anyway. Our friends are great and I received all sorts of compliments on both recipes but on the way home our ten-year-old, Lizzie, began asking me why it was such a big deal that "both of Liesa's dishes had wheat in them." She couldn't understand why that would matter and why people would even notice. I told her cooking with wheat is a bit unusual and tends to stand out. Lizzie just nodded her head a little, still perplexed, and then mentioned that her friend didn't seem to notice anything *unusual* about the sloppy joe she was gobbling up.

See, Lizzie's only a pup but she's grown up on wheat so she's not afraid of it. She doesn't know that for years and years it's been scoffed at even in Mormon communities where it's *always* been the very first item on any suggested food storage list. Lizzie's grown up hearing all the nutritional talk about wheat and the endless health benefits that come from incorporating it into our diets. She thought everyone was eating wheat all the time. Crazy kid.

Rice, beans, pasta, and potatoes—the other four basic dry goods we store, and eat—are more pedestrian. I can't help smiling when people say, "You really eat your food storage?"—like I'm serving bizarre dinners to my family. No, it's mostly the same ingredients I bought for years but now plan and purchase in quantities that would last at least three months and count as my food storage. The bizarre part is that everyday food actually *can be* food storage.

Now, let's look at everyday recipes in a new way

Ingredients for the recipes below are separated into two columns. All the ingredients that can be *stored* will be in the right column (I think its "right" to store those ingredients.). I'm hoping you'll mostly notice what's on the right. The left column is where I list all the remaining fresh ingredients (That's what's "left" to purchase. Get it?).

Wheat

Wheat is easier than you think. The first thing you should know is that the simplest way to cook wheat is in a slow cooker overnight. Before going to bed, pour four cups of wheat into ten cups of water, season with a tablespoon or less of salt, cover, and cook on low for eight to ten hours. This should make at least twelve cups of cooked whole wheat "berries," which can then be stored in three or four freezer bags and frozen until needed.

Basic Cooked Wheat Berries

Storage Ingredients
4 c. raw whole wheat
10 c. water
1 T. salt

Oil a large (4 quart or larger) slow cooker and fill with wheat, water, and salt. Cover and cook on low all night, 8-10 hours. Cooked wheat may be bagged and stored in the refrigerator for at least a week or in the freezer for months.

Hot Chicken and Wheat Casserole

Fresh Ingredients
½ c. butter
1 c. celery, diced
1 c. onion, diced
1 c. green pepper, diced
2 T. butter

Storage Ingredients
6 c. cooked wheat berries
2 cans chicken chunks (12.5 oz.)
1 can cream of chicken soup
(10¾ oz.)
¾ c. mayonnaise
1 can sliced water chestnuts (6 oz.)
1 c. breadcrumbs

Sauté diced fresh vegetables in ½ c. butter until they are partially cooked and color brightens. Add wheat berries, drained chicken chunks, soup, mayonnaise, and drained water chestnuts. Stir to combine. Pour mixture into an oiled 9" x 13" pan and sprinkle top with breadcrumbs that have been sautéed in 2 T. butter and lightly toasted. Bake uncovered at 350° for 45 minutes. This baked chicken salad casserole is one that actually tastes better if prepared a day ahead of time, making it an excellent option for entertaining on the run or for more peaceful Sunday dinners.

Once I served this as the main course for a baby shower at my sister's house, and weeks later someone hunted down my phone number and called me from out of state just to get the recipe. I can't remember ever working that hard to track down a recipe. I do remember that when the shower was over, my very cool high-school-age nephew and his basketball buddies literally cleaned out every bit of the leftover casserole.

Curry Chicken and Wheat Salad

Fresh Ingredients
4 c. diced cooked chicken
1½ c. celery, thinly sliced
2 c. red grapes

Storage Ingredients
3 c. cooked wheat berries
½ c. chopped walnuts
½ c. golden raisins

Dressing
¼ c. evaporated canned milk
1 c. mayonnaise
½ c. apricot preserves
¼ c. rice wine vinegar
2 T. curry powder
½ tsp. salt
½ tsp. pepper

Blend dressing ingredients and gently toss with salad ingredients. Refrigerate for at least 1 hour before serving. Serve over mixed greens or inside a fresh croissant.

When we first moved back to Salt Lake City, we had a family night picnic on the lawn of the state capitol. It was a beautiful warm night, and we could see over the whole valley as we ate our chicken salad and watched the sunset. When a daughter asked if it was okay for us to be there, my husband and I assured her that we were fine and that it was "our" state capitol. Then, near the end of our dinner we noticed a policeman walking up the big hill toward us. Well, it was getting late anyway, so we casually but quickly loaded up our supplies and headed for the car. Ahh yes, a FHE just barely on this side of the law.

Broccoli Wheat Salad

Fresh Ingredients
2 large broccoli, 8 c. chopped,
    florets only
1 c. Monterey jack cheese,
    grated
½ c. red onion, diced
½ lb. bacon, cooked
    and crumbled

Storage Ingredients
2 c. cooked wheat berries
½ c. Miracle Whip
½ c. mayonnaise
1 T. sugar

Toss all ingredients. Cover and chill, at least overnight, for the best combination of flavors.

This is our girls' favorite green salad. It's sort of a heartier spinach salad. The crumbled bacon, grated cheese, and slightly sweet dressing really make it. I served this when we invited Elder and Sister Wirthlin for Sunday dinner. It was such a treat to have them in our home, and they both seemed to enjoy this unusual salad. Always so thoughtful and kind, Elisa surprised me a few days later when she showed up at my door with a plate of her homemade chocolate chip cookies, prepared with whole-wheat flour. She knew where I was coming from. What a blessing to have known her. When I make this salad I remember Elisa.

Fiesta Wheat and Bean Salad

Fresh Ingredients
1 green pepper, diced
1 orange or red pepper, diced
1 small red onion, diced
1 bunch fresh cilantro,
   leaves chopped
1 small jalapeno pepper,
   cored and diced
juice of 1 lime

Storage Ingredients
3 c. cooked wheat berries
1 can corn, drained (16 oz.)
1 can black beans, drained
   (15 oz.)
2 t. chili powder
2 t. cumin powder
¼ c. cider vinegar
1 T. sugar
½ c. vegetable oil
1 t. salt

Combine all ingredients in a bowl. Toss well. Chill for at least 1 hour before serving.

Here's another salad that gets rave reviews. I've served it in groups that would laugh at the idea of eating your food storage, yet have had these same people call and ask me to bring this dish to the next function. Everyone wants to know if the brightly colored peppers are mangos. It's very fresh and mildly spicy and can be served as a salad, side dish, or a wonderful salsa for tortilla chips. A friend even uses the leftovers for days as the filling in simple veggie tacos. My favorite way is to grill cheese quesadillas, using white corn tortillas, and then serve them topped with the salad.

Sa-Wheat Sloppy Joes

Fresh Ingredients
1 lb. ground beef

Storage Ingredients
2 c. cooked wheat berries
1 can tomato soup (10¾ oz.)
1 jar Homade brand chili sauce
(12 oz.)

Brown the ground beef and drain fat. Stir in wheat, undiluted tomato soup, and chili sauce. Serve on toasted hamburger buns, with melted cheese if you like.

This is the perfect meal for soccer night, dance night, piano night, work night, or any night you need to have dinner waiting in the crockpot so that people can eat at different times. My sister Virginia has made this recipe for years without the wheat, and when I began adding it the change was imperceptible. Now that my daughters are older they dig the idea that there's fifty percent less red meat in each sandwich and one hundred percent more whole grain. I'm not exactly sure of that math, but you get the picture. The trickiest part of this recipe is finding the right chili sauce. I use a brand called "Homade Chili Sauce." It's sold in a little round glass jar about the size of a baseball. Keep in mind that if you were unable to obtain ground beef, the chili sauce and tomato soup could be added to four cups of cooked wheat berries for a great vegetarian-style emergency dinner.

Fabo Wheat Chili

Fresh Ingredients	Storage Ingredients
1 lb. ground beef	2 c. cooked wheat berries
1 onion, diced	2 cans tomato soup (10¾ oz.)
1 green pepper, diced	1 can diced tomatoes (14½ oz.)
	1 can kidney beans with liquid (1 lb. 11 oz.)
	½ tsp. paprika
	1 T. chili powder
	½ t. dry minced garlic
	1 t. salt
	¼ t. pepper
	1 bay leaf

Brown ground beef with diced onion and green pepper. Add remainder of the ingredients and simmer for 15 minutes. Garnish with sour cream and grated cheese. A single recipe yields 10 cups of chili, which serves 4–6 people. I usually make this in the early afternoon and then transfer it to a slow cooker, where it can simmer on low until we are ready to eat. This recipe also freezes well.

The secret ingredient here is the canned tomato soup. This recipe was found by my friend Cynthia and has won ward chili cook-offs and satisfied the finicky Young Women on numerous occasions. It's mild enough for children and rich enough for adults. The fact that there are whole-wheat berries added to the hamburger is hardly noticeable.

Wheat and Rice Bowls

Fresh Ingredients
4–6 green onions, thinly
    sliced or chopped
any combination of leftover
    vegetables: celery, carrots,
    peppers, peas, and corn
2 scrambled eggs (optional)

Storage Ingredients
1 can Spam, or leftover meat
3 c. cooked cold wheat berries
3 c. cooked cold rice
1–2 T. soy sauce
1 t. salt
¼ t. pepper
¼ c. oil
½ t. sugar

Dice Spam and sauté until browned and crispy. Set aside. If vegetables have not been cooked you may want to sauté them in the oil until tender. Then combine all ingredients in pan and heat thoroughly before serving. The ratio of wheat and rice can be adjusted to fit whatever is left in your refrigerator. A splash of teriyaki sauce is nice.

Stay with me—Spam is better than you would think when it's sautéed. And this recipe is a great way to use up all the leftover grains and veggies in the fridge. The simple recipe makes a ton, so be aware of that when you choose your pan. I'm not kidding. Invariably somehow this creates more leftovers, but they microwave well and actually get eaten. It can be seasoned with teriyaki or soy sauce, and you can use any leftover meats depending on your family's preference. I usually try to prepare this recipe in the early morning when I can sneak the time to get everything chopped and sautéed. Then it's ready, waiting for lunches or dinners, as the day and kids get busier.

Blender Wheat Pancakes

Fresh Ingredients	Storage Ingredients
1½ c. buttermilk	1 c. uncooked whole wheat kernels
2 eggs	2 T. cornmeal, heaping
2 T. butter, melted	1 T. brown sugar
	½ t. baking soda
	1 T. baking powder

Heat an electric frying pan or griddle to medium heat. Pour buttermilk into blender. Add the wheat kernels and blend on high for 10 minutes. It will sound like popping corn at first but will gradually become quieter. Add cornmeal, brown sugar, eggs, and melted butter, and blend another minute. Then add baking soda and baking powder and blend only a few seconds longer. Stir batter a little by hand to completely combine ingredients. Spoon onto hot griddle.

"Breakfast for dinner"… with no guilt. This one uses the uncooked wheat kernels right out of the can, is a great one for the kids to prepare, and helps them become comfortable with eating wheat. It also saves money in the weekly food budget when it's served as a dinner entrée, and who has time to make pancakes from scratch before school, anyway?

Rice

Anything served with rice is secretly a food storage meal. I'm forcing myself to include these recipes that use rice: it's hard for me to believe that anyone needs help knowing what to cook with rice, but maybe these ideas will be different from ones you've seen before or remind you of dishes you've cooked in the past. Mostly they are just easy recipes that our family has enjoyed forever. The reason I've selected them as part of our everyday food storage plan is that they are made with some canned, bottled, or dry ingredients, aren't too complicated, and go well with rice. Best of all, they help me quickly throw together "normal" food storage meals without having to run to the grocery store again and again.

Oven Rice

Fresh Ingredients
½ c. butter, sliced

Storage Ingredients
3 c. rice
6 c. water
6 chicken bouillon cubes
2 T. dry minced onion

Oil a 9" x 13" pan and add the rice and all other ingredients. Carefully stir. Cover tightly with foil. Bake at 325° for 1 hour. When done, fluff and mix with a fork before serving.

Usually I just cook rice on the stove; but sometimes when I need a large quantity, like for a party, I'll use this easy recipe for yummy oven-baked rice. I don't own a rice cooker, so this is my secret for great rice that I don't have to watch or worry about. I like that it takes two minutes to throw together and then it's into the oven for an hour. I like that I don't have to check on it or fuss over it. It always turns out great and the pan is easy to clean.

Beef Borgie

Fresh Ingredients
1 round steak cut into
   strips or cubes*

Storage Ingredients
*may substitute with 2 cans
   beef chunks (12 oz. each)
1 can cream of mushroom soup
   (10¾ oz.)
1 envelope Lipton Onion Soup
   mix
1 can sliced mushrooms (12 oz.)
2 T. flour
1 can evaporated milk (12 oz.)

Distribute pieces of round steak in an oiled 9" x 13" pan. Mix cream of mushroom soup, Lipton onion soup mix, and mushrooms with liquid; pour over beef. Cover, and bake at 300° for 3 hours. Stir in evaporated milk before serving. If substituting canned meat the ingredients can be heated on the stove and ready to serve before the rice is cooked.

I was given this recipe at one of my own wedding showers! Who knew I would still be using it as part of my food storage design twenty-five years later?

Swiss Chicken

Fresh Ingredients	Storage Ingredients
4 boneless, skinless chicken breasts	2 cans cream of chicken soup (10¾ oz.)
10 oz. Swiss cheese, sliced	1 can apple juice (5.5 oz.)
4 c. mushrooms, quartered	2 c. Pepperidge Farm Stuffing Mix
½ c. butter	

Cut each breast into five or six pieces and arrange in an oiled 9" x 13" pan. Cover chicken with a layer of sliced Swiss cheese and then a layer of mushrooms. Stir undiluted soup with apple juice and spoon over mushrooms. Sprinkle the top with crushed stuffing mix and drizzle with melted butter. Bake uncovered at 350° for 1 hour. Serve with hot rice.

This is a great recipe. It's fast and easy to prepare, and turns out to be a yummy chicken meal that everyone seems to enjoy. I serve it with hot rice, frozen string beans sautéed in a little olive oil with minced garlic, and a side of fresh tomato wedges.

Chicken and Broccoli

Fresh Ingredients
2 lbs. fresh broccoli,
    steamed crisp tender
½ c. grated cheddar cheese
½ c. grated Monterrey Jack
    cheese
2 T. butter

Storage Ingredients
2 cans cooked diced chicken
    (12.5 oz.)
1 can cream of chicken soup
    (10¾ oz.)
1 can evaporated milk (12 oz.)
½ c. mayonnaise
2 t. lemon pepper
1 t. curry
½ c. corn flakes or bread crumbs

In a 9" x 13" inch oiled baking pan, evenly arrange the steamed broccoli pieces. Mix drained chicken, cheeses, soup, milk, mayonnaise, lemon pepper, and curry. Spoon over broccoli. Top with crushed corn flakes or bread crumbs that have been sautéed and lightly toasted in butter. Bake at 350° for 40 minutes, the first 20 minutes covered with foil and the second 20 minutes uncovered. Serve with hot rice.

I use to poach the fresh chicken breasts and then dice them, before I discovered canned chicken. Trust me, no one can tell the difference if you take the short cut in this recipe. My family actually prefers the canned meat to the freshly cooked, insisting it's more tender (they're sort of rude that way).

Train Wrecks

Fresh Ingredients
Toppings:
green onion, diced
celery, diced
peppers, diced
carrots, shredded
cheese, grated

Storage Ingredients
Chicken Gravy:
2 cans cooked diced chicken
    (12.5 oz.)
1 can cream of chicken soup
    (10¾ oz.)
1 can evaporated milk (12 oz.)
2 cubes chicken bullion
¼ t. basil
½ t. salt
½ t. pepper

Toppings From Storage:
1 c. golden raisins
1 c. shredded coconut
1 can pineapple tidbits (14 oz.)
1 c. slivered almonds, toasted

Heat all chicken gravy ingredients, including liquid from the canned chicken. Prepare the fresh toppings listed, substituting with anything else you enjoy or have on hand. Serve buffet-style, beginning with a serving of hot rice, a scoop of chicken gravy, and everything else just piled on top.

Most people call these Hawaiian Haystacks. It's everything you can think of, layered over a bed of hot rice and chicken gravy. The hardest part is getting a few vegetables diced, cheese grated, and a can of pineapple drained. Then you have a refreshing and satisfying dish that's as fun to eat as a banana split for dinner. Almost.

Spam Bean-O

Fresh Ingredients
2 c. grated cheddar cheese
1 c. sour cream

Storage Ingredients
1 can Spam Luncheon Meat (12 oz.)
4 cans cut green beans (14.5 oz.)
1 can cream of mushroom soup
   (10¾ oz.)
¼ c. dry minced onion
salt and pepper

Dice Spam into ¼" cubes and sauté until browned and crispy on the edges. Drain green beans and pour into a 9" x 13" oiled pan. Sprinkle with salt and pepper, minced onion, and sautéed meat. Mix soup with sour cream and cheese; spoon on top. Bake uncovered at 350° for 30–40 minutes, until hot and bubbly. Serve over hot rice.

Yes, I really did say the S-word again. Here's the deal. When the Spam is diced into tiny chunks and sautéed until browned and slightly crisp like bacon, the flavor is wonderful and very much like bacon. We just don't tell people what they're eating. I made it once for my neighbor, Jeanette, to take to her gourmet Thanksgiving meal. After the dinner they have a discussion about what people really enjoyed out of the new recipes and this was chosen as a must-have for the next year. If you prefer, substitute canned ham.

Coconut Beef Curry

Fresh Ingredients
cilantro, chopped

Storage Ingredients
1 can beef chunks (12 oz.)
1 can coconut milk (13.5 oz.)
1 can diced potatoes (15 oz.)
1 can sliced carrots (15 oz.)
1 can peas (15 oz.)
2 T. flour
4 t. curry powder
¼ t. ground red pepper
1 envelope Lipton Onion Soup mix

In a medium-sized pot, combine beef chunks with broth, coconut milk, drained vegetables, flour, and spices. Simmer for 10 minutes and serve over hot rice, garnished with fresh chopped cilantro.

This one kicks. Canned vegetables aren't my favorite but they're smart to have in a food storage, and this recipe completely disguises them. Plus, I think it's cool to always have coconut milk on hand. Truly, this is curry in a hurry.

Easy Honey Mustard Chicken

Fresh Ingredients	Storage Ingredients
4 boneless, skinless, chicken breasts	⅓ c. honey
⅓ c. butter	⅓ c. prepared mustard

Place chicken in bottom of an oiled slow cooker. Mix together melted butter, honey, and mustard, pouring over top of chicken. Cover and cook on low for 6 to 8 hours. Serve with hot rice. This dish may also be baked in the oven at 350° for 1 hour if you are short on time.

I like this recipe because it cooks in the Crock-Pot, tastes great, and calls for minimal ingredients that store well.

Beans

I didn't know beans about beans. Turns out they're really very easy to cook but it does take a little time. I keep ten cases of dried beans purchased at the cannery in my long-term storage. I also buy a much smaller amount of cooked, canned beans from the grocery store. Depending on my work and family's schedule, I like having both bean options available.

Throughout this section of bean recipes, the canned amounts listed may be substituted with home-cooked beans from dry storage. For instance, if the recipe calls for a can of beans that weighs around 16 oz. then you should plan to soak and cook ⅔ cups dry beans for 2 cups of finished beans as a substitution. When using dried beans, the desired cooked amount is three times the dried amount. For example, ⅓ cup dried beans = 1 cup cooked beans. Below you'll find two basic recipes for cooking beans.

Basic Bean Recipe

Storage Ingredients
2 c. beans (1 lb. dry)
6 c. water

Sort and rinse beans. Bring 1 pound of beans to boil for 2 minutes in 8 cups of water, cover, and set aside to soak for 1 hour. Drain water and rinse beans. Add 6 cups of fresh water and simmer beans for 1½–2 hours.

If you would like to experiment with preparing just a small, one-pound package of dried beans from the grocery store, a similar recipe should be printed right on the back of each package.

Slow-Cooked Beans

Storage Ingredients
2 c. beans (1 lb. dry)
6 c. water

Sort and rinse beans. Cover 1 pound of beans with 6 cups of water and allow to soak for 24 hours. During this soaking process the water should be changed twice. Drain, rinse, and transfer beans to a slow cooker with 6 cups of fresh water. Cook on low for approximately 8 hours. The time will depend on the age of the beans. My dried beans are about ten years old and they still cook up just fine.

This second recipe for cooking beans works especially well if your dried beans have already been stored for a few years. The trick of pre-soaking older beans and then using a Crock-Pot to slowly cook them over many hours works wonders. Obviously this process requires planning ahead, but don't let that discourage you. The cooked beans can then be stored in the refrigerator or freezer until needed.

Taco Salad

Fresh Ingredients
lettuce
cheese
tomatoes
onions
sour cream

Storage Ingredients
canned chili
tortilla chips
salsa

Heat canned chili. Layer salad on individual plates beginning with lettuce, tortilla chips, warm chili, and then grated cheese. Continue to garnish as desired.

I already admitted to using canned chili for my taco salads. I find that if I buy different varieties and mix them the end result is ... less offensive, slightly better. I know that doesn't sound like great marketing but the point of this quick dinner is that it beats most frozen dinners in taste and nutritional value and the kids—even big kids—seem to enjoy it.

Big Bean Soup

Fresh Ingredients
1 onion, diced
1½ c. celery, diced
1½ c. carrots, diced

Storage Ingredients
2 c. pinto or white beans
   (pre-soaked 24 hrs)
6 chicken bouillon cubes
1 t. dry minced garlic
2 t. onion salt
1 T. crushed marjoram
⅛ c. Worcestershire sauce
⅛ t. pepper
1 can Spam, diced and sautéed
   until edges crisp (can substitute 3–4 pieces ham shank)

After beans have soaked, cook them in 8 cups of fresh water until soft, either on the stove or overnight in a slow cooker. The time will depend on your beans, so it's good to start early. Add remaining ingredients. Soup should simmer about an hour, until all the vegetables are fully cooked. Serve and enjoy!

My friend Mary Lynne taught me how to make this home-style soup that's very satisfying. My recipe calls for Spam, but I usually prepare it using three or four pieces of ham shank. I simmer them with the beans on low in my Crock-Pot, until the meat is just falling off the bones and the beans are tender. I still want to have Spam in my storage just so I have that meat option available. If Spam is sautéed and browned it really does taste like a nice cross between ham and bacon.

Taco Soup

Fresh Ingredients
1 lb. hamburger
1 onion, diced
1 c. celery, diced

Storage Ingredients
2 cans petite diced tomatoes
 (28 oz. ea.)
2 cans kidney beans (1 lb. 11 oz. ea)
1 can tomato juice (46 oz.)
1 can corn (15 oz.)
1 can cut green beans (15 oz.)
1 c. dehydrated carrot
1 pkg. taco seasoning mix

Brown hamburger with the diced onion and celery. Add all canned ingredients, including liquid; add dehydrated carrot, and stir in seasoning mix. Heat through on stove and serve with corn chips, grated cheese, sour cream, and a little fresh, chopped cilantro if desired.

My daughter Rosie ate this soup at her friend's house and came home to tell me, "Mom! You HAVE TO get the recipe from WenDee." I think the only thing I added was the dehydrated carrot. Any reason to add more instant vegetables to my food storage is a good one.

Baked Beans

Fresh Ingredients
1 onion, diced
1 green pepper, diced
1 lb. bacon, diced

Storage Ingredients
1 can Bush's Baked Beans (28 oz.)
1 c. barbeque sauce
2 cans black beans, drained (15 oz. ea.)
2 cans kidney beans, drained (15 oz. ea.)

Drain liquid from the three varieties of canned beans or any combination of cooked equivalent. Cook the diced bacon, draining off most of the drippings, and add the onion and green pepper. Continue cooking until the vegetables are lightly softened, and add all the beans and barbeque sauce. Add more sauce if needed. Simmer on low heat for 30 minutes.

When I find Bush's Best Baked Beans on sale for $1 per 28 oz. can I buy a couple of cases. I serve them with hot dogs and count it as a food storage dinner. If I want to take baked beans to a family party or barbeque, these few tricks dress them up nicely.

Split Pea Soup

Fresh Ingredients	Storage Ingredients
1 c. onion, diced	1 lb. dry green split peas (2 c.)
1 c. celery, diced	1 t. salt
2 c. potatoes, diced	½ t. pepper
2 c. carrots, diced	1 can Spam, diced and sautéed until crispy
	1 can evaporated milk (12 oz.)

Rinse and sort peas. Cook peas in 8 cups of water, in a slow cooker, set on high for five hours. Add diced vegetables, salt, and pepper, and continue cooking for 2 hours. When peas and vegetables are softened add the diced Spam that's been sautéed until the edges are crispy. Finish by adding the canned milk and serve.

Even if someone says they don't like split pea soup you should have them taste this recipe. I've served it at bridal showers and Relief Society dinners and been told more than once that it's the best split pea soup ever.

White Bean Chicken Chili

Fresh Ingredients	Storage Ingredients
1 c. onion, diced	2 c. white beans (pre-soaked for 24 hrs.)
1 c. celery, diced	2 cans chicken (12.5 oz.)
3 T. fresh lime juice	3 cans corn (11 oz.)
(optional)	2 cans chopped green chilies (4 oz.)
	2 t. lemon pepper
	2 t. ground cumin
	1 t. dry minced garlic
	8 chicken bouillon cubes

After beans have soaked, cook them until soft in 8 cups of fresh water, either on the stove or overnight in a slow cooker. Once the beans are fully cooked and soft, add all the other ingredients, including liquid, and simmer until onion and celery are tender. Serve with tortilla chips and top with a little grated Monterey Jack cheese, cilantro, and sour cream.

This is basically another recipe for taco soup but a white bean version that uses chicken instead of hamburger. Our family loves Mexican food and this is a long-time favorite.

Chicken Enchiladas

Fresh Ingredients
½ onion, diced and
   sautéed in ¼ c. butter
1 c. chopped cilantro
10 8" tortillas
2 c. grated cheese

Storage Ingredients
1 can chicken, drained (12.5 oz.)
1 can white beans, drained (15 oz.)
1 can cream of chicken soup
   (10.5 oz.)
1 can mild diced green chilies
   (7 oz.)
1 can evaporated milk (12 oz.)
1 c. leftover cooked rice

Oil a 9" x 13" pan. Drain the canned chicken and beans and mix them with the soup, green chilies, ¼ cup evaporated milk, cilantro and rice. Dip each tortilla in the remaining evaporated milk and fill with ½ cup of meat mixture. Roll tortillas enchilada-style and place in pan with seam down. Top with grated cheese. Cover and bake at 350° for 30 minutes. Uncover and bake an additional 5–10 minutes until hot and bubbly.

Here's an easy recipe for comfort food that's quick to assemble, using leftover rice and mostly canned ingredients. I feel like I'm beating the system when I make these enchiladas.

Refried Bean Burritos

Fresh Ingredients

tortillas

cheese

Storage Ingredients

2 cups instant refried bean flakes

2 cups boiling water

Boil water and stir in bean flakes. Cover and allow to sit for ten minutes. Fill tortillas with hot refried beans and top with grated cheese. Roll and cover with more grated cheese if desired. Microwave 1 minute. Garnish with sour cream, tomatoes, and cilantro.

I've taught my ten-year-old, Lizzie, how to prepare these burritos and she can make them for her friends. She used to beg me to buy those nasty little frozen pockets and pizzas all the time, but I wouldn't because of all the "who knows what" they contain. Now she can make her own instant lunch and spice it up however she likes. The refried bean flakes are usually available at the cannery.

Pasta

Did you know that macaroni stores a long, long time? When canned with an oxygen packet, the suggested shelf life is thirty years. I don't can my spaghetti noodles (we don't eat that much spaghetti) so they're instead part of my three-month storage, but macaroni is a good backup for any pasta dish.

Spaghetti

Storage Ingredients
1 jar spaghetti sauce
1 pkg. spaghetti

Cook the spaghetti following package directions. Heat sauce and serve over pasta. One pound of browned ground beef and fresh Parmesan cheese topping adds a lot.

Spaghetti was my very first food storage meal. I'm all about starting small and this is the easiest recipe on the planet! It's also a great place to begin teaching children how to cook for themselves. When it comes time for college, spaghetti is actually a life skill.

Mac & Cheese 101

Fresh Ingredients
4½ c. sharp Cheddar
   cheese, grated
¼ c. butter
3 slices of bread, crusts
   removed

Storage Ingredients
2 c. uncooked macaroni
2 cans evaporated milk (12 oz.)
¼ c. all-purpose flour
1 t. onion salt
⅛ t. nutmeg
⅛ t. pepper

Cook macaroni according to package directions; drain well. Set aside. Place milk, flour, onion salt, nutmeg, and pepper in a quart jar; cover tightly, and shake vigorously 1 minute. Stir together flour mixture, 3½ cups cheese, and macaroni. Pour into an oiled 9" x 13" pan and sprinkle with remaining 1 cup cheese and fresh breadcrumbs. Drizzle with melted butter and bake at 350° for 45 minutes.

I don't think there's anything wrong with buying cases of Kraft Macaroni and Cheese for food storage. That's exactly what our children preferred so that's what we used to store. Now that they're older they appreciate this easy-to-make homemade version.

Killer Fettuccine

Fresh Ingredients	Storage Ingredients
½ c. butter	¼ t. dry minced garlic
8 oz. cream cheese	¼ t. salt
1 c. whipping cream	1 pkg. fettuccine noodles

Slowly heat butter, cream cheese, whipping cream and seasonings. Pour over any cooked pasta and/or meat and vegetable combination.

As soon as you read the list of ingredients you'll understand the name of this recipe. As soon as you taste it you'll agree with me. When I was in high school I worked at a nice family-style restaurant where they served this cheese sauce on at least five different dishes. One of the best was over omelets, but it was also great as a dipping sauce on a steamed veggie plate or with diced chicken and pasta for a yummy fettuccini. After the cooks gave me the secret recipe I knew why all the customers just loved it. Mama Mia.

Chicken Noodle Soup

Fresh Ingredients
1 onion, diced
1 c. celery, diced

Storage Ingredients
2 cans cooked chicken chunks (12 oz.)
8 c. water
½ c. dehydrated carrot
6 cubes chicken bouillon
1 can cream of chicken soup (10¾ oz.)
1 can evaporated milk (12 oz.)
2 c. pasta

Everything except the pasta may be cooked on low in a slow cooker for 3–4 hours. If you're short on time the vegetables can be microwaved and then added. When almost ready to serve, add the pasta of your choice and simmer another 20 to 30 minutes. Extra seasonings such as parsley, rosemary, and fresh ground pepper may also be added.

There are two different ways that I make chicken noodle soup. Both cook in the slow cooker but one method uses a whole fresh chicken, cooks all day long, and takes some time to remove all the bones. The other, more food-storage-friendly way, uses cans of cooked chicken chunks and is fairly instant. They both have their place, but I'm sharing the food storage recipe here.

Mato Mac

Fresh Ingredients
¼ onion, diced
¼ celery, diced

Storage Ingredients
1 c. macaroni
1 can tomato juice (46 oz.)
¼ c. dehydrated carrot
1 can roast beef chunks with liquid
    (12 oz.)
1 can evaporated milk (12 oz.)

If adding diced, raw vegetables, first zap them in the microwave to speed cooking time. Simmer all ingredients in tomato juice until fully cooked. Salt and pepper to taste.

My friend Melanie taught me that you can cook a cup of macaroni with a large can of tomato juice, add a little salt and pepper, and have a very simple tomato macaroni soup. I like to add a can of roast beef chunks, a little dehydrated carrot, canned milk, and whatever leftover vegetables happen to be in the refrigerator. Either way, it's a quickie dinner that tastes great.

Tuna Noodles

Fresh Ingredients
1 c. cheese, cubed
  or shredded
1 c. frozen peas

Storage Ingredients
2 c. macaroni or noodles
1 can cream of mushroom soup
  (10¾ oz.)
1 can evaporated milk (12 oz.)
2 cans tuna, drained (6 oz.)

Boil macaroni in salted water for 7 minutes and drain. Stir soup, milk, peas, tuna and pasta together in a 2-quart casserole. Bake at 400° for 20 minutes, or until hot. Top with cheese.

This dish reminds me of my mom. It was one of the quick and easy ways she fed six children. I still think of it as comfort food.

Beef and Veggie Noodle Bowl

Fresh Ingredients
3 c. broccoli florets
3 medium carrots,
   peeled, sliced (2 c.)
1 red pepper, diced
1 lb. beef sirloin steak,
   cut into strips

Storage Ingredients
8 oz. linguine, uncooked
2 t. oil
¼ c. KRAFT Zesty Italian Dressing
¼ c. teriyaki sauce
1 t. ground ginger
½ c. dry-roasted peanuts

Cook pasta as directed on package, adding vegetables to the cooking water for the last 2 minutes of pasta cooking time. Drain pasta mixture. Meanwhile, heat oil in large nonstick skillet. Add meat; cook until browned on both sides. Stir in dressing, teriyaki sauce, and ginger; cook until sauce thickens slightly. Toss pasta and meat mixtures in large serving bowl and top with peanuts.

I love Asian food but don't like to buy the twenty-five different spices required in some of those wonderful sauces. This recipe comes together quickly, has the kick I'm craving, and uses basic ingredients.

Pesto Pasta

Fresh Ingredients
1½ c. basil pesto

Storage Ingredients
2 c. macaroni

Boil macaroni in 5 cups of salted water for 7 minutes. Drain. Stir in basil pesto, or any other variety of jarred pasta sauce. Serve with possible additional items such as tomato wedges, a few pieces of roasted chicken, or a little fresh cream drizzled on top.

This is another family favorite because it's so simple and really good. I buy the basil pesto in the deli section at Costco. The twelve and under set seem to prefer the pasta plain, which works great for me.

Potatoes

Instant potatoes are like fast food in the world of food storage. But they can store for thirty years.

BBQ Chicken

Fresh Ingredients
4–6 boneless, skinless,
    chicken breasts

Storage Ingredients
16 oz. barbeque sauce
1 T. lemon pepper

Rinse chicken and place in oiled slow cooker. Sprinkle with lemon pepper and cover with barbeque sauce. Cook on low heat 4–5 hours. Serve with instant mashed potatoes.

Nothing is easier than putting a few chicken breasts in a slow cooker and covering them with BBQ sauce. This dish is easy to prepare as we're walking out the door on Sunday morning, and ready to eat as soon as we get home. I usually serve my BBQ chicken with mashed potatoes because I really want to be using the instant potatoes I have stored.

Italian Chicken

Fresh Ingredients
4 chicken breasts*

Storage Ingredients
*substitute with 2–3 cans cooked
    chicken chunks
1 can cream of chicken soup (10½ oz.)
1 can evaporated milk (12 oz.)
1 pkg. Italian dressing seasoning

Oil slow cooker and add thawed, boneless and skinless chicken breasts. Combine soup and dry salad dressing seasoning and pour over chicken. Cook on low for 4–5 hours. When ready to serve, stir canned milk into the sauce. Serve over instant mashed potatoes.

This is a surprisingly light recipe for tender chicken in a creamy white sauce. Fresh vegetables like onions, carrots, celery, and peppers may also be added to the Crock-Pot, cooking along with the chicken to create a rich stew that is wonderful served over mashed potatoes, hot noodles, or plain rice.

Shepherd's Pie

Fresh Ingredients
1 lb. ground beef*
grated cheese

Storage Ingredients
*substitute with 1–2 cans cooked
  beef chunks
1 can tomato soup (10¾ oz.)
2 cans cut green beans (15 oz.)
instant potatoes

Prepare instant potatoes so that you have about 4 cups ready. Brown ground beef and drain fat. (If substituting with canned beef chunks, drain liquid.) Add undiluted tomato soup to beef. Drain green beans and pour into a large pie pan. Cover with meat and tomato soup mixture and top with prepared mashed potatoes. Sprinkle with grated cheese and bake uncovered at 350° for 30 minutes.

Didn't everyone learn how to cook this in junior high cooking class? I did. I still use it as part of my regular family meals because it's simple, satisfying, and uses ingredients from food storage.

Meatloaf

Fresh Ingredients
1 lb. ground beef
2 eggs, beaten
¼ c. onion, diced

Storage Ingredients
⅔ c. regular dry oats or breadcrumbs
¾ c. evaporated canned milk
1 t. salt
⅛ t. pepper
½ t. ground sage

Glaze:
¼ c. ketchup
1 t. prepared mustard
¼ t. nutmeg
3 T. brown sugar

Mix all meat loaf ingredients together and shape into a 9" glass pie plate. Mix glaze ingredients and spread over top of meat loaf. Cover with plastic wrap and microwave on full power for 10 minutes. Add additional time for desired doneness. Serve with instant mashed potatoes.

I use at least three different meatloaf recipes but I'm sharing this one because it cooks in the microwave and has a fun tangy glaze on top.

Cranberry Pork Chops

Fresh Ingredients
4–6 thick pork chops

Storage Ingredients
1 bottle French dressing (8 oz.)
1 can whole cranberries (14 oz.)
1 envelope Lipton Onion Soup Mix

Brown the pork chops on the stove in a little olive oil, then transfer to an oiled 9" x 13" baking pan. Mix together the dressing, cranberries, and onion soup mix. Pour over pork chops and bake at 350° for 1 hour. Serve with hot instant mashed potatoes.

I like having cranberry sauce in my food storage for variety, and the blend of sweet and spicy flavors pair well with pork. Another good option is to buy the cannery's pear sauce when it's available and keep it on hand to serve with basic pork chops or roasted pork tenderloin.

Fast Funeral Potatoes

Fresh Ingredients
¼ c. chopped green onion
½ c. shredded cheese

Storage Ingredients
6 c. prepared instant potatoes
1 can cream of chicken soup
   (10¾ oz.)
1 can evaporated canned milk
   (12 oz.)

Topping
2 T. melted butter
½ c. shredded cheese

1 c. crushed cornflakes

Spoon prepared mashed potatoes into an oiled 9" x 13" baking dish. Mix soup, canned milk, green onions, and shredded cheese, and pour over potatoes. Gently fork liquid mixture into potatoes. Stir topping ingredients (melted butter, cheese, and cornflakes) and distribute over potatoes. Bake at 350° for 30 minutes.

Who knows how many versions there are of this classic, "cheesy" potato recipe. A quick version is prepared using instant mashed potatoes instead of frozen hash browns or shredded boiled potatoes.

Harvest Potato Soup

Fresh Ingredients
1 c. onion, diced
1 c. celery, diced
4 c. potato, diced
2 c. grated cheese
½ lb. bacon, cooked
   and diced

Storage Ingredients
2 c. instant potatoes
½ c. dehydrated carrot
8 chicken bouillon cubes
1 can cream of chicken soup
   (10¾ oz.)
1 can evaporated milk
   (12 oz.)

Simmer all vegetables, bouillon cubes, and cream of chicken soup in 8 cups of water until vegetables are fully cooked. Add instant potatoes and evaporated milk to achieve desired consistency. Garnish with grated cheese and bacon bits.

Here's a little soup invention that uses up more of the leftovers in the refrigerator. It's not so much a recipe as an idea. Consider combinations your family would like, and experiment with your favorite meats and vegetables.

Tomato Artichoke Soup

Fresh Ingredients
4 T. butter
1 lg. onion, finely chopped
1 T. minced garlic
1 c. sour cream

Storage Ingredients
1 can petite diced tomatoes
   with juices (28 oz.)
4 chicken bouillon cubes
4 c. water
1 can artichoke hearts, coarsely
   chopped (15 oz.)
½ c. dehydrated carrots
1 t. dried thyme leaves
½ t. salt
¼ t. pepper
¼ t. baking soda
1 T. sugar

In a large saucepan, heat butter and sauté onion for 5 minutes, until soft. Add garlic and thyme, sautéing another minute. Add the tomatoes, bouillon cubes, water, artichoke hearts with their juices, dehydrated carrots, salt, pepper, baking soda, and sugar. Cover and simmer for 30 minutes. Add the sour cream, stir well and simmer for 5 more minutes.

I love this soup because it's another recipe that uses mostly canned ingredients from my food storage and is still gourmet enough to be served at parties. (I always double the recipe.) You may notice that there are no instant potatoes in this recipe. Sorry about that. It's my book.

Oats

Canned rolled oats have an estimated shelf life of thirty years. We have ten cases of oats in our long-term storage, and even though I don't use them in dinner recipes, I work through my supply by using them in cookies, apple crisp, and breakfast cereals. I prefer the regular variety over quick oats because I want minimal processing.

Berry Wonderful Breakfast Cobbler

Fresh Ingredients
½ c. frozen blueberries

Storage Ingredients
¼ c. rolled oats (old fashioned or non instant)
¼ c. walnuts, crumbled
1 T. golden raisins
1 T. Craisins
1 T. brown sugar
½ t. cinnamon
½ c. water

Toss all ingredients except blueberries in a large cereal bowl and microwave for 2 minutes. Add frozen blueberries and microwave an additional 1½ minutes. Add milk and enjoy!

Recently a friend tasted this at my house and asked me how I prepare it for the whole family. I responded like a deer in the headlights—I have to admit I've never made it for the whole family. We have a strong "self-serve" breakfast policy. Maybe that's a good thing. I don't know. The advantage to this recipe is that you make it, as needed, one bowl at a time.

It's a breakfast I look forward to and I'm not even a fan of oatmeal. I read about health benefits of blueberries, oatmeal, walnuts, golden raisins, cranberries, and cinnamon, but the reason I eat this practically every morning is that it tastes like dessert. That's the way I want to start my day!

Cranberry Granola

Storage Ingredients
½ c. oil
½ c. pure maple syrup
1½ c. brown sugar
6 c. oats
1 c. chopped almonds
1 c. chopped pecans
1 c. wheat germ
1 c. sweetened shredded coconut
1 c. currants or golden raisins
1 c. dried cranberries

Place oven racks in the two center positions. Preheat oven to 350°. Spray two jelly roll pans with cooking oil spray. Set aside.

In medium bowl, combine oil, syrup and brown sugar. Microwave uncovered for 3 minutes or until sugar starts to melt. Remove from microwave and whisk until any lumps disappear.

In large bowl, combine dry ingredients. Toss to mix well. Pour syrup mixture over dry mixture and stir until well mixed. Spread evenly onto jellyroll pans.

Place both pans in oven and bake for 10 minutes. Remove pans from oven, stir granola, then place back in the oven, rotating to the opposite rack. Bake for 8 minutes more before removing from oven. Cool to room temperature. Stir again. Store in airtight container up to one month. Makes about 4 quarts.

My neighbor, Bonnie, gave me some of this granola in a gift bag as a Christmas present, and I loved it! Way better than anything you can buy in a box, and it's a snap to prepare. If people in your family prefer cold cereal, this is the recipe for you.

GET GOING

Moving Beyond Food Storage Myths

More and more people are recognizing the advantages of provident living, and that eating your food storage actually makes life easier in daily routines as well as emergency situations. It's wonderful that the general climate surrounding food storage is beginning to warm up. If you're still feeling chill to the concept, but hopefully working on overcoming objections, this next chapter will help as a reality check and pep talk.

If you're going to have a food storage, you'll probably need to cook

I have great friends who insist they can't cook and have no desire to learn. Some even argue that it's a luxury they're entitled to because of their busy careers. I relate to that sentiment because I'm pretty busy myself and don't enjoy cooking. During my freshman year at BYU my roommate, Julie, and I took an actual photograph of the one meal we prepared. Those were some amazing omelets and I still have the picture to prove it! I seriously don't know how we survived living on our own. I do remember lots of angry notes taped to kitchen cupboards and refrigerator shelves where our roommates demanded that we "DO NOT TOUCH!" any of their food.

I should have learned a lot more from my mother. Even though I grew up in a home where everything was made from scratch, I rarely helped in the kitchen. While she was making dinner I was at dance lesson, play practice, or out with friends. Nowadays I cook mostly to get the job done. I care about trying to cook well because it serves my family and friends. Even though it's not one of my gifts, food preparation is a significant part of the life job I have chosen. It's one way I "do" for others and is constantly needed.

Outsourcing most food needs works as long as nothing ever changes in our lives or our society, but you know what they say about change. It just doesn't make sense to deliberately limit ourselves and the people around us by not having a basic knowledge of how to prepare food. Even if it's a bitter pill and even if you're only looking to provide for yourself, food preparation is still a life sustaining skill, helps every individual become a better provider, and is worth the effort it takes to learn. No one should settle for being merely a consumer. That's just too limiting.

And, it may be time to get real about canned food

Our neighbors prepared an amazing meal for us, where they served herb-encrusted pork tenderloin, perfectly roasted root vegetables, and spring greens with fresh grapefruit and berries in a light vinaigrette dressing. It was food heaven and we never wanted to leave. On the way home my daughter Rosie exclaimed that it was her favorite meal *ever* and I wholeheartedly agreed. Like most people, I would naturally prefer to only eat the very freshest fruits, vegetables, and meats, but I don't let that preference get in the way of having viable food storage.

A friend once confessed that she couldn't bear to eat canned fruits and vegetables. "That's okay!" (I didn't know what else to say to her. I was thinking that she probably needed to buy a farm near San Diego, tomorrow, but I kept it to myself.) We each have to come to terms with our wants and needs. The lifestyle that goes with food storage isn't the fanciest. Canned fruits and vegetables, and cream of mushroom soup, aren't cool gourmet ingredients but they're helpful tools, and that's good enough for me. The blessings of provident living simply matter more than the farmer's market.

You may want to evaluate how you're using the storage space in your home

Food storage takes space, and that's something we rarely have enough of. I live in a home that was built before closets were invented, so I've had to get creative with our food storage. It's meant storing food under beds, but more important than that, it's meant letting the less essential things go to make room for supplies that could be life saving. My friend Stephanie recently told me she was going to throw out miscellaneous items she had carefully stored for years in order to have more space for food storage. Food has to be more valuable than endless Halloween costumes and old toys.

At the same time, you shouldn't have to add on to the back of your house to have space for your food storage. Long-term food storage for a family of five, consisting of seven stacks lined up next to each other—ten cases of wheat, ten cases of rice, ten cases of beans, ten cases of pasta, ten cases of potatoes, ten cases of oats, and an additional ten cases of wheat (again, because it's so inexpensive and lasts the longest)—will only be about 7' wide x 6' high x 19" deep.

Minimal space required for long-term food storage

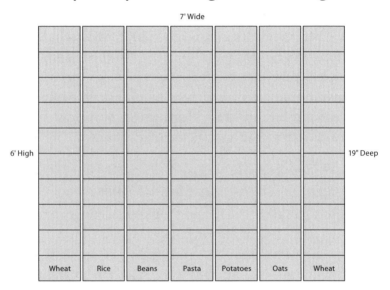

7' Wide

6' High

19" Deep

Wheat | Rice | Beans | Pasta | Potatoes | Oats | Wheat

That's like a large bookcase or wall unit that is completely modular, can be reconfigured as needed, will be useful for the next thirty years or longer, and has the potential to save lives. Compared to all the other things we buy for our homes, this kind of "storage unit," which costs around $1,200, is a smart investment and worth much more than the space it occupies.

You don't have to be rich

Preparedness isn't just a plan for the wealthy. Don't be discouraged if you have very little to spend. Heavenly Father perfectly understands each individual situation, and He doesn't compare one person's offering of obedience to another's. He sees what we choose in life and He knows our hearts. The example from the scriptures is that a widow's mite was more valuable to Him than all that the wealthy could give. I believe He feels the same way about one can of food. If you buy one can of anything for food storage, it counts!

One Christmas each of our four older girls received a $50 check from their Great Aunt Mary. Teenagers, poised to blow the cash on CDs and earrings, were asked to contribute to the family food storage budget. *Not* initially in love with this idea, they complained but listened as we pointed out that for only $50 they could buy into a wonderful program, offering improved safety for themselves and others. So much more than a $200 infusion for our storage budget, we wanted each daughter to experience the peace that comes from investing in the Lord's way of doing things.

We had friends in New York who could afford to make a phone call and within days a semi-truck with plenty of manpower was unloading a complete year's supply of food into their storage room. That was their opportunity and a wonderful blessing. Isn't it great they chose to do that with their money? Of course it is, but I don't really believe that the Lord values one wealthy family's great offering any more than the person who struggles to just get by and yet willingly makes sacrifices in order to give their best, no matter how humble their offering of obedience. I believe we're judged mercifully as individuals and never in comparison to each other.

The First Presidency acknowledges that some are strictly limited in their ability to prepare:

> We realize that some of you may not have financial resources or space for such storage. Some of you may be prohibited by law from storing large amounts of food. We encourage you to store as much as circumstances allow. May the Lord bless you in your home storage efforts. ("All Is Safely Gathered In: Family Home Storage," Intellectual Reserve, Inc., 2007)

And if you are rich, there are "abundant options" to consider

I've heard accounts where after a disaster a car gets traded for a bike and then the bike gets traded for a bowl of rice within an hour. My friend Khaliel told me of her epiphany when she saw a newspaper photo of men wearing Rolex watches standing in line for food during a disaster. Bottom line for me is that some of us have been blessed to enjoy more of the world's abundance. If that's your situation, and to the degree that it's mine, I suggest we consider amassing some wealth—like stores of extra food—which truly does satisfy. Eatable wealth can sustain life. I suggest there will be the greatest comfort, enjoyment, and happiness in serving family and neighbors as true humanitarians.

If you bought your food storage 20–30 years ago,
it might be time to recycle and rebuild

We've offered our newly married daughter first dibs on
hand-me-down food storage in slightly outdated cans
of beans, wheat, and rice, as she and her husband are
ramping up their own family budget. The products may
not be "first quality" but would provide a viable safety
net—certainly better than nothing—and bridge the gap
until they can afford to purchase their own. I've also tak-
en the opportunity to offer food storage to the missionaries and various people I work with. Passing it along to
someone else is just another way to share some of the
wealth I've been blessed with.

Recently I was at a meeting with our Stake Presidency when the question was asked, "What should be
done with the giant drums of ancient wheat we've had
stored in the basement, for who knows how long, and
never opened?" "Donate it, or chuck it, then buy more"
was my gentle suggestion. I'm betting that anyone able
to afford buying food storage many years ago can probably afford to buy a fresh supply of the inexpensive basic
grains again today, and enjoy the advantages of #10 cans
in neat and manageable cases. If people are currently en-
joying life with a few day-to-day luxuries, wouldn't they
want to re-invest in the luxury of food storage they can
actually plan to eat, share, and enjoy? The old drums
are proof that they believe in listening to the prophet, so
if there's sufficient money in the "golden years" budget
it doesn't make sense to scrimp by with the "moldy-oldy"
or simply stop listening. Food storage is one area of the
gospel where enduring to the end doesn't have to be a
huge challenge.

What if we never use our food storage?

Over the years I've heard people make comments like, "When my mother died we had to throw out cases of old wheat she had stored under her bed." I'm not sure what their point is. Do they wish she had lived to *have to* use her storage? Or do they wish their mother hadn't obeyed the prophet? Maybe the blessing that followed obedience was that she *didn't* have to use her food storage.

Those comments make me think that what's really being said is, "I've paid my insurance premiums for years and dang it if not one of my kids has ever been hit by a car!" I don't understand the frustration. Waste is bad and we should all try to limit that as much as we can, but being caught without food is really, really bad. In our society most of us waste money virtually every day. What about the last Big Gulp we bought, or the lame movie, or too expensive sweater? Those things happen almost weekly for some and while I'm not suggesting it's okay to waste food storage, I think we can keep it in perspective with other financial decisions. I know I've wasted money on far less worthy efforts.

There are a few things I would suggest you try, and try to avoid

As you've seen from the recipes I use, I've tried to keep things focused on what is simple and fits the needs of my family. There are seminars on food storage given by people who are making their own cheese, and that's not especially useful to me. While I respect that skill as a wonderful talent, I personally just can't dive that deep. Cooking with powdered eggs and making every single thing from scratch is too intimidating and overwhelming, and I honestly don't have time. Instead, I've really appreciated that the most current word on food storage counsels us to "Build a small supply of food that is part of your normal, daily diet." How liberating to know that it's okay to begin exactly where we are.

Food is very personal. I'm uncomfortable with huge expanded lists or programs from the past that tell me exactly what I should buy at the store. It's too much information, and I don't trust that whoever is behind the list somehow knows how much cornmeal and instant pudding I will need in a year. It's like having someone else buy your swimsuit. That would *never* work. When it comes to food I'm really the only one who can even hope to know what will fit my family.

I also don't like the idea of buying a year's supply of dehydrated camping food. First of all, my family doesn't eat that way even when we're camping, and all that I've ever tasted is a bit too much like cat food. I don't know a single person who ever eats their dehydrated food. The rhythm seems to be that it gets bought, stored in the garage, eventually turns into something else, and then gets tossed. For sure it's better than nothing, but it's just too odd and too expensive for me.

Try developing some new skills and get ready to give

A major motivation for me is the ability to take care of my family. I want to improve on my mommy skills; I don't want to fail the people I love. I have spent years refining my professional craft and spent all sorts of money on an expensive education. I know the Lord is aware of how hard I have worked to develop those groups of talents. I don't want to be caught in a personal or community disaster and not have useful skills to share with my family. Chances are, in a real emergency, no one is going to need a new sofa.

So what can I offer my family, my friends, and the people in my neighborhood? In Proverbs 31:20 we read: "She stretched out her hand to the poor: yea, she reacheth forth her hands to the needy." I want to be able to help. I don't want to run to the bishop and tell him I forgot to prepare. I don't want to do that to my family. And I don't want to have to learn how to cook funky foods during a crisis. Even if I feel I could be effective under pressure, why would I bet on that?

Get prepared to comfort the children

Speaking of pressure, I want to do all I can, while we enjoy relative peace, to minimize the possible future pressure on my children. I feel that having to say, "Okay kids, we're in an emergency, everything you know is gone or has drastically shifted, and now we're going to start eating really strange food," isn't very kind. My plan is that if I can learn now, in peace and safety, how to prepare my food storage in normal ways that my family actually enjoys, in the event of an emergency I can easily modify or simplify recipes and prepare things that would be *familiar, recognizable*, and hopefully *satisfying*. I'm thinking of food storage as "long-term comfort food" that can be shared with my family and my neighbors.

Years ago I read the book, *I Walked to Zion: True Stories of Young Pioneers on the Mormon Trail* (Salt Lake City, UT: Shadow Mountain Pub., 1994). Every pioneer account was moving, but the one I've never forgotten was from a child who actually *enjoyed* the trek across the prairie. The journey was described as a magical adventure, and the simple reason this child enjoyed crossing the plains was because the mother of the family had somehow secured a small portable oven before setting out and was therefore able to bake fresh bread all along the way.

I learned two things from that story. Number one: I seriously needed to learn how to bake bread (and I did). Number two and perhaps more important: the preparation and skill of a mother can completely alter the experience of a child. Choosing to prepare food storage and developing my ability to cook with it will make me better able to gather those I love and say with confidence, "We're going to get through this." Children glean strength from their parents, and if I have done all I can to prepare, developing useful skills and drawing my strength from the Lord and his prophets, I'm going to be in a better position to help and support others.

Teach them to fish

I told my girls that if they would learn how to grind wheat and make bread, completing the entire process ten times by themselves, I would buy them a Bosch mixer and electric wheat grinder when they married. It's an idea I gained from a wonderful homemaker and teacher, Ann Allred. Following her example, I named this incentive program the Liesa Card Scholarship, and one daughter took the bait. Using the bread-making experience as one of her Laurel projects, Sarah was surprised at how easy the process was and even shared her new talent with a few of her friends. Later, when she attended BYU, her roommates loved it when their apartment filled with the aroma of baking bread.

Yes, it takes a wad of cash to buy a quality bread mixer and grinder, but I want Sarah to have the best tools available so that her bread making can be efficient and successful. I'm so glad she didn't wait until she was thirty-something to learn, like I did. It always bothered me that women across the world with far less education and no modern advantages have been baking bread for thousands of years and yet today it's practically become a lost art. Making bread is a life-sustaining skill.

I'm trying to learn and teach my children new skills. Most of our children are adults now and have moved away from home, but my husband and I encourage them to implement the principles of preparedness: just as they need to pay their own tithing, they need to take responsibility for their own preparedness if they want the attendant blessings. Provident living is best taught through actions, rather than lessons. I have no idea whether food storage will bless my life primarily as an excellent organizational tool, but I pray that my children will see my obedience, know that I believed in following prophets, and make their own choices, helping themselves and future generations to prepare and serve others.

We sent a few cans of rice and oats with our eldest when she left for BYU. A couple of years later, we mailed a $50 check to our daughter attending Boston University and asked her to purchase her own emergency supplies—bottled water, flashlight and batteries, granola bars, rain poncho, band aids, extra socks, and gloves—that could be kept in her dorm room. We also sent $50 with our daughter on study abroad in Romania, asking her to similarly prepare during the semester she spent serving in hospitals and orphanages of that country. Our fourth daughter spent the last couple of years working as a professional model, living in Singapore, New York, Milan, and Tokyo. No matter how jam-packed her suitcases are with clothes and shoes, we make sure she leaves home with a couple of gallon Ziploc bags full of basic emergency supplies.

I can't protect my girls from everything that could go wrong in the world, but I still want to help and encourage them to prepare. So far, no one has had occasion to use her emergency supplies, but as concerned parents we feel peace in knowing we're assisting our children when they're far from home. We talk about the blessings of personal preparedness, try to demonstrate a good example, and then ask them to do all they can to follow the prophet's counsel to create an emergency storage. Beyond that, we pray that each one will stay close to the Lord and follow the promptings of the Holy Ghost.

TRUST

Conclusion: Preceding Testimony with Action

Within weeks of formally beginning work on this book, my seventy-nine-year-old mother suffered a stroke. The last ten years of her life have been troubled by the effects of Alzheimer's disease, and this stroke has left her with even greater complications and needs. Luckily for me, and with the much-needed assistance from my siblings and family, Marilyn Bennett Romney has been able to live with us. During this writing process, I sit at my computer, and she's sound asleep on a bed right behind me: there's something meaningful in that arrangement.

My mother's life has always been right behind me, as a constant example that gives strength to my journey. Her faithfulness proved a profound commitment to following prophets. Provident living wasn't merely a nice suggestion; it was the way she lived. And she wasn't only preparing for stormy weather or financial reversal. She prepared every day for eternity. During the good times and in trials, I could see that my mother had a testimony of following the prophet and clearly knew the smart way to go.

I don't know a single person who has a burning testimony and love for the temple but rarely goes. I only know people who love the temple and have a deep appreciation for it *because* they regularly attend and make room in their life for temple service. Testimonies aren't cheap. They come to us only after consistent obedience.

A testimony of food storage is similar. We learn in the doing. When I've taken a friend to the cannery for the first time they have left feeling surprised at how efficient and well thought out the process is. They're surprised to have felt the Spirit there. See this hand waving wildly? Me too! And yet growing up surrounded by good examples and having the opportunity to eat from my parents' food storage, I didn't actually come to my own testimony of it as being vital and founded on the principle of love until after I had taken my own steps to be obedient. If you've never felt the burn, or have let your testimony grow cold, just start moving and see if things don't heat up for you. I predict with full confidence that the Spirit will let you know you're on the right track.

I tell my daughters that faith is a verb. It means we *do* something. And often we have to take steps in the dark. Faith wouldn't be a test if we could clearly see all the answers and the future laid out, in the bright sunshine, beaming in front of us. Faith means we move forward with trust, carefully listening and trying to follow the guidance we've been given, and praying all along the way. Each person who buys an extra can of food is exercising faith. We can't see what's ahead—the when, how, and why we need food storage—but we can still follow inspired leadership.

This past Christmas I gave each of my daughters a large cellophane bag holding food for a 72-hour kit. On the outside of each bag I wrote what my sister Carolyn calls the "all that" promise found in 3 Nephi 1:13. Notice that this scripture is the voice of the Lord directly answering an "exceedingly sorrowful" (v. 10) Nephi who had "cried mightily unto the Lord all that day" (v. 12) in behalf of faithful people who were on the verge of immediate destruction:

> Lift up your head and be of good cheer; for behold, the time is at hand, and on this night shall the sign be given, and on the morrow come I into the world, to show unto the world that I will fulfill *all that* [emphasis mine] which I have caused to be spoken by the mouth of my holy prophets. (3 Nephi 1:13)

Food storage comes to us by the mouth of holy prophets. The promise given in this scripture, that the Lord "will fulfill all that which [He has] caused to be spoken by the mouth of [His] holy prophets" comforts, encourages, and motivates us to try to listen better and follow more closely the counsel that we've been given. We can completely trust the prophet's counsel because his words are backed by Jesus Christ. There is everything to be hopeful about: we really can "lift up [our] heads and be of good cheer." The great plan of happiness is a plan for success, and there's enough room for everyone.

Resources

Cost and Weight Recommendations for Long-term Storage

The Church's Provident Living website suggests that each adult needs 25 lbs. of wheat, white rice and other grains, plus an additional 5 lbs. of dry beans and other legumes, per month (www.providentliving.org). Using the food storage calculator on the website you can easily determine your family's food storage needs.

Besides knowing what and how much to buy, it seems that most people, including myself, *really* want to know *how big* their family's one year supply will be and *what it will cost them* to purchase. I've created the following chart to illustrate the cost of a basic, long-term storage. Please keep in mind that this amount of food, which is enough for five adults, is about 7' wide, 6' high, and 19" deep, and costs around $1,200.

A similar supply may be adjusted to suit any size of family. As described in Chapter 3, simply count the number of people and multiply it by two: the number you get will tell you the number of cases you'll purchase in each of the seven groups of long-term storage items. For example, based on this formula a family of three (3 x 2 = 6) might buy six cases of wheat, six cases of rice, six cases of beans, six cases of pasta, six cases of potatoes, six cases of oats, and an additional case of wheat (or whatever else is preferred) to obtain the total amount in grains and beans suggested by the Church.

Cost and Weight Example for 5 People and One Year			
Wheat, hard white	10 Cases	348 lbs.	$142.20
Rice, white	10 Cases	342 lbs.	$185.40
Bean, pinto	10 Cases	312 lbs.	$208.20
Pasta, macaroni	10 Cases	204 lbs.	$204.00
Potato flakes	10 Cases	108 lbs.	$175.80
Oats, regular	10 Cases	162 lbs.	$120.00
Wheat, hard white	10 Cases	348 lbs.	$142.20
Total	**70 Cases**	**1,824 lbs.**	**$1,177.80**

Card Family Trivia

Our family prefers hard white wheat because of its milder flavor. We buy the dried pinto beans because they are the least expensive variety, and then we buy additional grocery store cans of cooked white, black, and kidney beans as part of our three-month supply. We use only macaroni for our year supply of pasta and then buy three packages of spaghetti, fettuccini, and egg noodles for our three-month supply. Home Storage Centers now sell potato flakes as well as potato pearls. The potato pearls are a little more expensive, have butter and salt added, come *pre-packaged* which is a terrific advantage, but have a limited storage life. (My potato pearls are from January 1999 and they are in perfect condition. I also know people whose pearls didn't last the two years they were on a mission, so go figure.) The potato flakes are less expensive, unseasoned, require canning, and have a storage life of 30 years.

A current price list of products available from Family Home Storage Centers can be found at the website www.providentliving.org.

Food Storage Bulletin Boards for Meal Planning

One thing I've done to ensure that I don't get lost in the process of cycling through all thirty everyday meals — because I take restaurant and leftover breaks every now and again—is to create food storage bulletin boards with all of my recipes on display.

Really, it does help. They're actually just bigger, enhanced versions of the planner list I carry with me. Like, good-looking recipe maps, the bulletin boards track progress made toward preparing meals from storage. Mounted to the inside of a cupboard, attached to a shelving system, or simply hung on the wall, they're "in-your-face" support for *eating your food storage.*

Ingredients are printed on the front of each recipe card, and the cooking instructions are printed on the back. As each food storage meal is prepared, the recipe card is simply turned over and left in that position. Recipe cards facing front, with the ingredients showing, signal meals that still need to be used. Cards facing backwards, with the cooking instructions showing, like check marks on my planner list signal meals that have already been used.

There's something about attractive organizational systems that make them more likely to be used. Just ask Restoration Hardware, Pottery Barn, and West Elm. Those companies are doing a big business in color-coordinated boxes, bins, and bulletin boards. Food storage bulletin boards are great big reminders of what "*can*" be (sorry) for dinner, and having most of the necessary ingredients already on the shelf is a great incentive for cooking and eating more family meals at home.

Secretly, or not, with these added steps in support, I'm also hoping that my daughters will be better empowered to jump into easy meal preparation and experiment with cooking using their food storage. As they gain confidence and competence they're better prepared for being the woman in charge.

If these food storage bulletin boards sound helpful, you'll find the instructions below.

I have five of them—one each for wheat, beans, rice, pasta, and potatoes. (I can't believe I'm teaching a craft. Never say never.)

Step 1: Form the backing

A. Start with a 30" x 20" sheet of ³⁄₁₆" thick foam board. If you want to make five boards (one for each long-term storage food category) you will need two sheets of foam board.

B. Using a craft knife, and a new blade, cut each foam board into three 10" x 20" rectangles.

C. On the back of the foam board, measure carefully and then draw two long vertical guidelines 2½" in from both sides.

D. Mark a dot on both guidelines, measuring from one end to 1½", 7¾", and 14" (see photo).

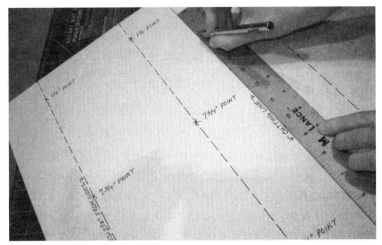

Mark guidelines on foam board

Step 2: Cover foam board with fabric

A. Select a fabric you like, something that adds interest to your kitchen, and cut to approximately 13" x 23."

B. Wrap the fabric evenly around the board and secure with masking tape, so that you can still see the marking dots on the back of the foam board.

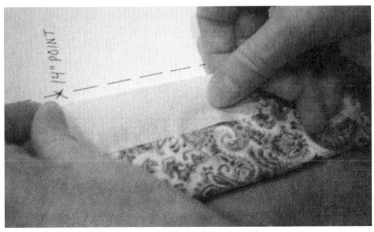

Secure fabric to foam board

Step 3: Construct recipe hangers

A. Thread a craft needle with a long piece of very heavy-duty thread or fine string.

B. Knot the end of one thread, leaving about a 3" tail.

C. Sew buttons on the front side of the bulletin board, so that your recipes will have something to hang on.

Note: If you don't use a gentle touch with this sewing process your string can tear a hole in the foam board. If that happens, reinforce the torn area with a patch made from layers of masking tape and adjust your spot slightly to accommodate a new stitch.

1. Begin with your first dot marked at 1½" from the top and sew through the back side to the front side of the board. A thimble really helps with this step. You need to attach the button with only a single stitch and then stitch through the front side to the back of the board.

Attach button hangers

2. Without cutting the thread, move on to your next marked dot (like tying a quilt) and repeat the sewing step above to attach another button. When you come to the end of all six dots and have your six buttons sewn on the front, cut your remaining thread, leaving another 3" tail.

3. Cut the thread in the center, between each marking dot, and carefully tie into at least a double knot to secure each button to the face of your board. For a bit of added stability, cover each knot with a piece of masking tape.

When that minimal sewing step is completed you should have a highly attractive, lightweight, inexpensive, personally designed bulletin board with six decorative buttons on the front for displaying your everyday food storage meals. (P.S. For the record, I wanted to kill myself by this point but I forced myself to keep going. I hate crafts even more than cooking. Maybe the title for this book should be "Hated it. Did it anyway. Food storage. One woman's story.")

Step 4: Attach recipes

A. Begin with 3" x 4" clear plastic pockets normally used for creating hanging name badges (Avery Hanging Naming Badges, product #74459) and available at most office supply stores.

B. Thread the pre-punched holes with a 12" piece of tiny ⅛" ribbon and knot together, leaving 1½" tails (see photo).

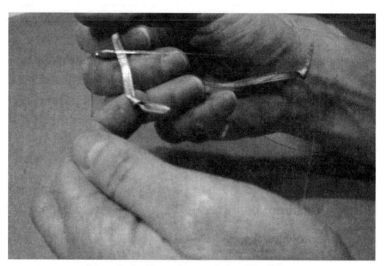

Thread plastic recipe pockets with ribbon

C. Print your recipes on cardstock cut to fit into the clear plastic pockets. List the ingredients on the front side and the cooking instructions on the back.

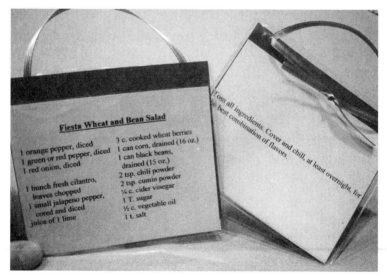

Insert recipes into plastic pockets

D. Hang six recipes on the front of the board.
    After you've prepared a meal, remember to turn over
    the recipe card so that the cooking instructions are
    visible, and then leave it in that position. This way,
    as soon as you open the cupboard you'll be reminded
    of what you've already prepared from your storage
    and which recipes are still waiting to be served.

Step 5: Mount to cupboard door or wall

The final step in using the bulletin boards is to mount
them inside the kitchen cupboard using a piece of adhe-
sive Velcro. If you have an entire storage room or a good-
sized pantry, the boards can be mounted to the wall or
the back of a door.

Mount food storage bulletin board to cupboard door

Sample Three-Month Storage Inventory and Shopping List Spreadsheet

The spreadsheet below offers an extended example of the three-month storage inventory and shopping list discussed in Chapter 3. It is organized by long-term storage food category and based on recipes from Chapter 4. The list can be used to check supplies and to plan shopping trips for storage and fresh ingredients.

Wheat Inventory and Shopping List				
Recipe				**3 Months**
Hot Chicken and Wheat Casserole	**Fresh Ingredients**	**Storage Ingredients**		
	½ c. Butter	2 cans	Chicken Chunks	6
	1 c. Celery	1 can	Cream of Chicken Soup 10¾ oz.	3
	1 c. Onion	0.75 cup	Mayonnaise	2.25
	1 c. Green Pepper	1 can	Sliced Water Chestnuts 6 oz.	3
	2 T. Butter	1 cup	Bread Crumbs	3

Fiesta Wheat and Bean Salad	**Fresh Ingredients**	**Storage Ingredients**		
	1 Orange Pepper	1 can	Corn 16 oz.	3
	1 Green Pepper	1 can	Black Beans 15 oz.	3
	1 Red Onion	2 tsp.	Chili Powder	6
	1 Bunch Cilantro	2 tsp.	Cumin Powder	6
	1 Small Jalapeno Pepper	0.25 cup	Cider Vinegar	0.75
	1 Lime	0.5 cup	Vegetable Oil	1.5

Wheat Inventory and Shopping List (continued)				
Recipe				**3 Months**
Sa-Wheat Sloppy Joes	**Fresh Ingredients**	**Storage Ingredients**		
	1 lb. Ground Beef	1 can	Tomato Soup 10¾ oz.	3
	1 c. Celery	1 jar	Homade Chili Sauce 12 oz.	3

Fabo Wheat Chili	**Fresh Ingredients**	**Storage Ingredients**		
	1 lb. Ground Beef	2 cans	Tomato Soup 10¾ oz.	6
	1 Onion	1 can	Petite Diced Tomatoes 14½ oz.	3
	1 Green Pepper	1 can	Kidney Beans 1 lb. 11 oz.	3
		0.5 tsp.	Paprika	1.5
		1 T.	Chili Powder	3
		0.5 tsp.	Dry Minced Garlic	1.5
		1	Bay Leaf	3

Curry Chicken and Wheat Salad	**Fresh Ingredients**	**Storage Ingredients**		
	4 Boneless Skinless Chicken Breasts	0.5 cups	Walnuts	1.5
	3 Ribs Celery	0.5 cups	Golden Raisins	1.5
	2 c. Red Grapes	1 can	Evaporated Milk	3
		1 cup	Mayonnaise	3
		0.5 cups	Apricot Preserves	1.5
		0.25 cups	Rice Wine Vinegar	0.75
		2 T.	Curry Powder	6

Wheat Inventory and Shopping List (continued)				
Recipe				**3 Months**
Broccoli Wheat Salad	**Fresh Ingredients**	**Storage Ingredients**		
	2 Lg. Broccoli, 8 c. Flowers Only	0.5 cups	Miracle Whip	1.5
	1 c. Monterey Jack Cheese	0.5 cups	Mayonnaise	1.5
	Red Onion			
	½ lb. Bacon			

Rice Inventory and Shopping List				
Recipe				**3 Months**
Beef Borgie	**Fresh Ingredients**	**Storage Ingredients**		
	2 lbs. Round Steak	2 cans	Beef Chunks 12 oz.	6
		1 can	Cream of Mushroom Soup 10¾ oz.	3
		0.5 box	Lipton Onion Soup 1 Envelope	1.5
		1 can	Evaporated Milk 12 oz.	3

Swiss Chicken	**Fresh Ingredients**	**Storage Ingredients**		
	4 Boneless Skinless Chicken Breasts	2 cans	Cream of Chicken Soup 10¾ oz.	6
	10 oz. Swiss Cheese	1 can	Apple Juice 5.5 oz.	3
	4 c. Sliced Mushrooms	0.33 bag	Pepperidge Farm Herb Stuffing	0.99
	½ c. Butter			

Rice Inventory and Shopping List (continued)

Recipe			3 Months
Chicken and Broccoli	**Fresh Ingredients**	**Storage Ingredients**	
	2 lbs. Fresh Broccoli	2 cans	Chicken Chunks 12 oz.	6
	½ c. Grated Cheddar Cheese	1 can	Cream of Chicken Soup 10¾ oz.	3
	½ c. Grated Monterey Jack Cheese	1 can	Evaporated Milk	3
	2 T. Butter	0.5 cup	Mayonnaise	1.5
		2 tsp.	Lemon Pepper	6
		1 tsp.	Curry Powder	3
		0.5 cup	Corn Flakes	1.5

Train Wrecks	**Fresh Ingredients**	**Storage Ingredients**	
	1 c. Diced Green Onion	2 cans	Chicken Chunks 12 oz.	6
	1 c. Diced Celery	1 can	Cream of Chicken Soup 10 3/4 oz.	3
	1 c. Diced Red Pepper	1 can	Evaporated Milk 12 oz.	3
	1 c. Shredded Carrots	2 cubes	Chicken Bouillon	6
	1 c. Grated Cheese	0.5 tsp.	Basil Leaves	1.5
		1 cup	Golden Raisins	3
		1 cup	Coconut, Shredded	3
		1 cup	Pineapple Tidbits 15 oz.	3
		1 cup	Slivered Almonds	3

Rice Inventory and Shopping List (continued)

Recipe				3 Months
Spam Bean-O	Fresh Ingredients	Storage Ingredients		
	2 c. Grated Cheese	1 can	Spam 12 oz.	3
	1 c. Sour Cream	4 cans	Cut Green Beans 14.5 oz.	12
		1 can	Cream of Mushroom Soup 10¾ oz.	3
		0.25 cup	Dry Minced Onion	0.75

Coconut Beef Curry	Fresh Ingredients	Storage Ingredients		
	Chopped Cilantro, Garnish	1 can	Beef Chunks 12 oz.	3
		1 can	Coconut Milk 13.5 oz.	3
		1 can	Diced Potatoes 15 oz.	3
		1 can	Sliced Carrots 15 oz.	3
		1 can	Peas 15 oz.	3
		4 tsp.	Curry Powder	12
		0.25 tsp.	Ground Red Pepper	0.75
		0.5 box	Lipton Onion Soup 1 Envelope	1.5

Bean Inventory and Shopping List

Recipe			3 Months
Taco Salad	Fresh Ingredients	Storage Ingredients	
	Lettuce	2 cans	Chili	6
	Cheese	0.33 bag	Tortilla Chips	0.99
	Tomatoes	1 jar	Salsa	3
	Onions			
	Sour Cream			

Big Bean Soup	Fresh Ingredients	Storage Ingredients	
	1 Onion	6 cubes	Chicken Bouillon	18
	1½ c. Celery	1 tsp.	Dry Minced Garlic	3
	1½ c. Carrots	2 tsp.	Onion Salt	6
		1 T.	Crushed Marjoram Leaves	3
		2 T.	Worcestershire Sauce	6
		1 can	Spam 12 oz.	3

Taco Soup	Fresh Ingredients	Storage Ingredients	
	1 lb. Ground Beef	2 cans	Petite Diced Tomatoes 28 oz.	6
	1 Onion	2 cans	Kidney Beans 1 lb. 11oz.	6
	1 c. Celery	1 can	Tomato Juice 46 oz.	3
		1 can	Corn 15 oz.	3
		1 can	Cut Green Beans 15 oz.	3
		1 can	Dehydrated Carrots	3
		1 pkg.	Taco Seasoning	3

Bean Inventory and Shopping List (continued)

Recipe				3 Months
Split Pea Soup	Fresh Ingredients	Storage Ingredients		
	1 c. Onion	1 lb. pkg.	Dry Green Split Peas	3
	1 c. Celery	1 can	Spam 12 oz.	3
	2 c. Potatoes	1 can	Evaporated Milk 12 oz.	3
	2 c. Carrots			

White Bean Chicken Chili	Fresh Ingredients	Storage Ingredients		
	1 c. Onion	2 cans	Chicken Chunks 12.5 oz.	6
	1 c. Celery	3 cans	Corn 11 oz.	9
	2 Limes	2 cans	Chopped Green Chilies 4oz.	6
		2 tsp.	Lemon Pepper	6
		2 tsp.	Group Cumin	6
		1 tsp.	Dry Minced Garlic	3
		8 cubes	Chicken Bouillon	24

Chicken Enchiladas	Fresh Ingredients	Storage Ingredients		
	½ Onion	1 can	Chicken Chunks 12.5 oz.	3
	¼ c. Butter	1 can	White Beans 15 oz.	3
	10 8" Tortillas	1 can	Cream of Chicken Soup 10¾ oz.	3
	2 c. Grated Cheese	1 can	Mild Diced Green Chilies 7 oz.	3
		1 can	Evaporated Milk 12 oz.	3

Pasta Inventory and Shopping List				
Recipe				**3 Months**
Spaghetti	**Fresh Ingredients**	**Storage Ingredients**		
	1 lb. Ground Beef	1 jar	Spaghetti Sauce	3
		1 pkg.	Spaghetti Noodles	3

Mac & Cheese 101	**Fresh Ingredients**	**Storage Ingredients**		
	4½ c. Sharp Cheddar Cheese	2 cans	Evaporated Canned Milk	6
	¼ c. Butter	0.25 cup	Flour	0.75
	3 Slices Bread	1 tsp.	Onion Salt	3
		0.12 tsp.	Nutmeg	0.36

Chicken Noodle Soup	**Fresh Ingredients**	**Storage Ingredients**		
	1 Onion	2 cans	Chicken Chunks 12 oz.	6
	1 c. Celery	0.5 cup	Dehydrated Carrot	1.5
		6 cubes	Chicken Bouillon	18
		1 can	Cream of Chicken Soup 10¾ oz.	3
		1 can	Evaporated Milk 12 oz.	3

Mato Mac	**Fresh Ingredients**	**Storage Ingredients**		
	Onion	1 can	Tomato Juice 46 oz.	3
	Celery	0.25 cup	Dehydrated Carrot	0.75
		1 can	Beef Chunks 12 oz.	3
		1 can	Evaporated Milk 12 oz.	3

Pasta Inventory and Shopping List (continued)

Recipe				3 Months
Tuna Noodles	Fresh Ingredients	Storage Ingredients		
	1 c. Cheese	1 can	Cream of Mushroom Soup 10¾ oz.	3
	1 c. Frozen Peas	1 can	Evaporated Milk 12 oz.	3
		2 cans	Tuna Fish 6 oz.	6

Beef and Veggie Noodle Bowl	Fresh Ingredients	Storage Ingredients		
	3 c. Broccoli	8 oz.	Linguine	24
	3 Carrots	2 tsp.	Olive Oil	6
	1 Red Pepper	0.25 cup	KRAFT Zesty Italian Dressing, Bottled	0.75
	1 lb. Beef Sirloin Steak	0.25 cup	Teriyaki Sauce	0.75
		1 tsp.	Ground Ginger	3
		0.5 cup	Dry Roasted Peanuts	1.5

Potato Inventory and Shopping List

Recipe			3 Months
B.B.Q. Chicken	**Fresh Ingredients**	**Storage Ingredients**	
	4–6 Boneless Skinless Chicken Breasts	1 bottle	B.B.Q. Sauce 16 oz.	3
		1 tsp.	Lemon Pepper	3

Italian Chicken	**Fresh Ingredients**	**Storage Ingredients**	
	4 Boneless Skinless Chicken Breasts	1 can	Cream of Chicken Soup 10¾ oz.	3
		1 can	Evaporated Milk 12 oz.	3
		1 pkg.	Italian Dressing Mix	3

Shepherd's Pie	**Fresh Ingredients**	**Storage Ingredients**	
	1 lb. Ground Beef	1 can	Tomato Soup 10¾ oz.	3
	1–2 c. Grated Cheese	2 cans	Cut Green Beans 15 oz.	6

Meat Loaf	**Fresh Ingredients**	**Storage Ingredients**	
	1 lb. Ground Beef	1 can	Evaporated Milk 12 oz.	3
	2 Eggs	0.5 Tsp.	Ground Sage	1.5
	¼ c. Onion	0.25 cup	Ketchup	0.75
		1 Tsp.	Prepared Mustard	3
		0.25 Tsp.	Nutmeg	0.75
		3 T.	Brown Sugar	9

Potato Inventory and Shopping List (continued)

Recipe			3 Months
Cranberry Pork Chops	**Fresh Ingredients**	**Storage Ingredients**	
	4–6 Thick Pork Chops	1 bottle	French Dressing 8 oz.	3
		1 can	Whole Cranberries 14 oz.	3
		0.5 Pkg.	Lipton Onion Soup	1.5

Fast Funeral Potatoes	**Fresh Ingredients**	**Storage Ingredients**	
	½ c. Green Onion	1 can	Cream of Chicken Soup 10 3/4 oz.	3
	1 c. Cheese	1 can	Evaporated Milk 12 oz.	3
	2 T. Butter	1 cup	Corn Flakes	3

Author photo: Duston Todd

Liesa Card has presented to numerous audiences on food storage topics, and she maintains a blog with continually updated food storage information, suggestions, and recipes at www.idareyoutoeatit.com. For the past 23 years she has also run her own successful interior design business. She and her husband, Mike, are the parents of five daughters and reside in Salt Lake City, Utah.